Luzia Götz

Applying a systematic conservation-planning tool with real data

Luzia Götz

Applying a systematic conservation-planning tool with real data

A first attempt

AV Akademikerverlag

Impressum / Imprint

Bibliografische Information der Deutschen Nationalbibliothek: Die Deutsche Nationalbibliothek verzeichnet diese Publikation in der Deutschen Nationalbibliografie; detaillierte bibliografische Daten sind im Internet über http://dnb.d-nb.de abrufbar. Alle in diesem Buch genannten Marken und Produktnamen unterliegen warenzeichen-, marken- oder patentrechtlichem Schutz bzw. sind Warenzeichen oder eingetragene Warenzeichen der jeweiligen Inhaber. Die Wiedergabe von Marken, Produktnamen, Gebrauchsnamen, Handelsnamen, Warenbezeichnungen u.s.w. in diesem Werk berechtigt auch ohne besondere Kennzeichnung nicht zu der Annahme, dass solche Namen im Sinne der Warenzeichen- und Markenschutzgesetzgebung als frei zu betrachten wären und daher von jedermann benutzt werden dürften.

Bibliographic information published by the Deutsche Nationalbibliothek: The Deutsche Nationalbibliothek lists this publication in the Deutsche Nationalbibliografie; detailed bibliographic data are available in the Internet at http://dnb.d-nb.de. Any brand names and product names mentioned in this book are subject to trademark, brand or patent protection and are trademarks or registered trademarks of their respective holders. The use of brand names, product names, common names, trade names, product descriptions etc. even without a particular marking in this work is in no way to be construed to mean that such names may be regarded as unrestricted in respect of trademark and brand protection legislation and could thus be used by anyone.

Coverbild / Cover image: www.ingimage.com

Verlag / Publisher:
AV Akademikerverlag
ist ein Imprint der / is a trademark of
OmniScriptum GmbH & Co. KG
Heinrich-Böcking-Str. 6-8, 66121 Saarbrücken, Deutschland / Germany
Email: info@akademikerverlag.de

Herstellung: siehe letzte Seite /
Printed at: see last page
ISBN: 978-3-639-62625-4

Content

1. Introduction

Taking conservation actions has not arisen recently. It is an old, established practice that has been applied in various cultures all over the world. More recently, conservation areas have been established in order to protect biodiversity and ecosystems. These areas should ensure that conservation features are separated from threats to their existence. But whether this strategy is effective depends on how well it responds to the subsequent two questions. First, how well do conservation areas represent the whole of biodiversity? Second, how well does a reserve systems support the long-term viability of species, ecosystems and the respecitve essential processes? At this point systematic conservation planning has attended to increase the ability of conservation actions to successfully respond to these questions (Margules and Pressey 2000). The field of systematic conservation planning has emerged over the past 30 years and has influenced planning processes of governmental administrations and non-governmental organisations (Moilanen, Wilson et al. 2009). At the same time, computer based planning tools have been developed and applied worldwide. However, in Switzerland systematic conservation-planning tools have been used only once (Bolliger, Edwards et al. 2011). Hence, this master thesis seeks to implement the concept of systematic conservation planning for a part of Switzerland by applying a systematic conservation-planning tool.

The software that is possibly the most widely used systematic conservation-planning tool in the world is Marxan. The objective of Marxan is to solve one type of the minimum set, reserve design problem. Thus, it implements an optimization algorithm that selects planning units based on user-defined targets at minimum cost (McDonnell, Possingham et al. 2002). I apply this optimization algorithm to real data on the species existing in conservation areas of cantonal interest in canton Aargau. My study attemps to illustrate several important issues relevant to implementing a systematic conservation-planning tool for an existing network of conservation areas. First, the study organizes appropriate species data for each of the conservation areas. Second, it includes a method of cost calculation that could be adopted for further analyses with Marxan. Third, it illustrates how Marxan selects the conservation areas with the given cost, the available species data and a variation of user-defined targets. Fourth, it discusses the inputs and outputs of Marxan in cosiderations of its application as a decision support tool for Swiss conservation practice. Finally, the study emphasizes the importance of further steps in the planning process in consideration of the key principles of systematic conservation planning.

2. Methods

2.1 Study Site

Canton Aargau is situated in the north of Switzerland and covers approximately 140'400ha in the Swiss lowlands and Jura. The landscape of this canton is diverse. One third of the area is covered with forest and, hence, there are about 4'000km of forest edges. Agriculture land covers about 44% and represents the main type of open land ecosystem. There are about 528km of hedges and tree lines that cross the open land. A small part of the area of canton Aargau, 2.6%, is defined as unproductive land. These are often wetlands that still exist along the rivers Reuss, Limmat and Aare. Canton Aargau has sought to protect such rare habitat types. In fact, it is the canton's obligation to take conservation actions. But this canton's engagement is above the ordinary. Today canton Aargau contains 376 conservation areas of cantonal interest (CCI). Most of these areas are located in open land; only 25 CCIs are in forest (Fig. 1). In the open land, CCIs are dry meadows and pastures, wetlands, bogs, reeds and even former gravel-pits. On the dry meadows and pastures, hedges often exist as well. Especially in the larger CCIs, different ecosystem types and ecological structures are located side by side. The CCIs cover a total area of 2'279ha and, thus, about 1.6% of the cantons area. The smallest CCI constitutes 0.14ha and the largest 109ha (Canton Aargau 2013).

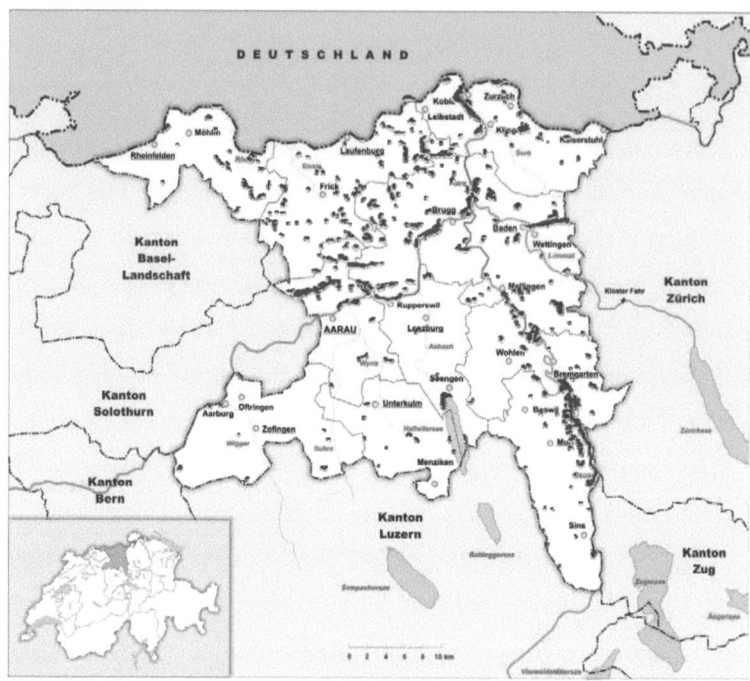

Figure 1: Conservation areas of cantonal interest (green) in canton Aargau

2.2 Systematic Conservation Planning

Taking conservation actions by humans has been an old-established procedure and has taken place worldwide. But protected areas used to be implemented because they were unusable for any commercial purpose or they were abandoned and remote sites. Conservation areas realized in this way don't achieve the two main roles of providing habitat for the biodiversity of each region and break it away from the processes that endangered its survival. The consequences are that all the species which have their habitat in productive areas, are not protected. Thus systematic methods for selecting and designing conservation areas are required (Margules and Pressey 2000).

The field of systematic conservation planning has been evolved for more than 30 years. During this time the term has been defined through several characteristics, principles and descriptions of the planning process. Margules and Pressey (2000) pointed out six distinctive characteristics of systematic conservation planning:

1. It requires clear choices about the features to be used as surrogates for overall biodiversity in the planning process.
2. It is based on explicit goals, preferably translated into quantitative, operational targets.
3. It recognizes the extent to which conservation goals have been met in existing reserves.
4. It uses simple, explicit methods for locating and designing new reserves to complement existing ones in achieving goals.
5. It applies explicit criteria for implementing conservation action on the ground, especially with respect to the scheduling of protective management when not all candidate areas can be secured at once (usually).
6. It adopts explicit objectives and mechanisms for maintaining the conditions within reserves that are required to foster the persistence of key natural features, together with monitoring of those features and adaptive management as required.

After this, Moilanen, Wilson et al. (2009) crystallized the key principles for spatial prioritization problems (Fig. 2). These principles have arisen with the increasing interest in technical approaches for conservation planning. They all address critical characteristics in the field of planning, designing and managing conservation reserve systems.

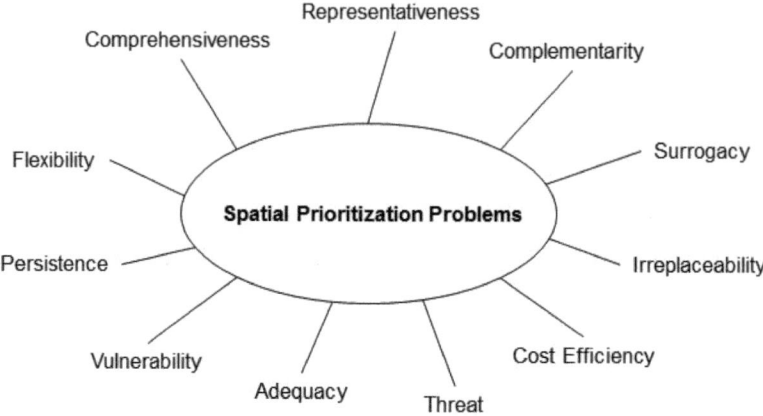

Figure 2: The key principles for spatial prioritization problems have arisen with the increasing interest on technical approaches for conservation planning. They all address critical characteristics in the field of planning, designing and managing conservation reserve systems.

Comprehensiveness and representativeness refer to a network of conservation areas that contains an assemblage of every biodiversity feature including considerations about composition, structure and function. A comprehensive and representative network of conservation areas could be accomplished with a set of complementary conservation areas. To overcome the common lack of data, numerous surrogates have been proposed (Ferrier 2002).The two principles adequacy and persistence point the adequacy of a network for ensuring the permanence of the protected biodiversity features. To comply with these principles considerations on future circumstances have to be taken. Cost efficiency is one principle that systematic conservation-planning tools especially address on. Costs are basic restrictions in taking conservation actions and thus couldn't be discounted. Regarding the target of minimizing biodiversity loss, the respective threats and vulnerability of

8

conservation features have to be incorporated. Further it is crucial to additionally account for future threats. If a planning unit is of particular importance for a reserve system it indicates irreplaceability. This could be the case if a planning unit contains a singular occurrence of a conservation feature. Finally, by allowing for alternative opportunities in combining planning units, flexibility grants finding a planning unit that has a similar adequacy.

There are several suggestions how to implement the key principles in the planning process. One suggestion of a planning process was the eight stages of systematic conservation planning that has been worked out from several sources (Ardron et al. 2010):

1. Identify and involve stakeholders
2. Identify goals and objectives
3. Compile Data
4. Establish conservation targets and design principles
5. Review existing protected areas and identify network gaps
6. Select new protected areas
7. Implement conservation action
8. Maintain and monitor the protected area network

This study focused especially on the planning stages 3, 4 and 5. Thus, it covered merely a part of the whole long time planning process.

2.3 Marxan

As analytical tool the decision-support software Marxan for systematic conservation planning was applied. Marxan functions to solve the 'minimum set problem', where user-defined targets are met at the lowest possible cost (McDonnell, Possingham et al. 2002, Game and Grantham 2008). In respect to limited resources, it implements the principle of

9

efficiency. To address this principle, Marxan works with different optimization algorithms that implement simulated annealing. Simulated annealing acts by selecting a set of planning units that meet the user-defined targets whereas the score is kept to a minimum (Pearce, Kirk et al. 2008). Targets are set for each species individually. It determines the number of individuals that are intended to exist in the final reserve system. When all the targets are met the score is as high as the sum of the costs of the selected planning unit. If Marxan can't find a reserve system that meets all the targets, the score consists of costs and penalties. The pentalty is determined by a species penalty factor (SPF). This factor is a multiplier and could be defined by the user. The penalty is greater the higher the value of the SPF is set. The optimization process takes place several times (determined by the number of runs) where the current solution is randomly replaced by a more close-by optimal solution. Especially in the beginning of the optimization process with simulated annealing bad moves are stochastically allowed. Through the stochastic way of the simulated annealing optimization algorithm, it avoids getting stuck at local minima (Fig. 3; Ardron et al. 2010). It is obvious that the optimization algorithm would get stuck at the indicated local minimum if the stochastic jumps to the right wouldn't be allowed. This is the key feature of the simulated annealing optimization algorithm in comparison to others.

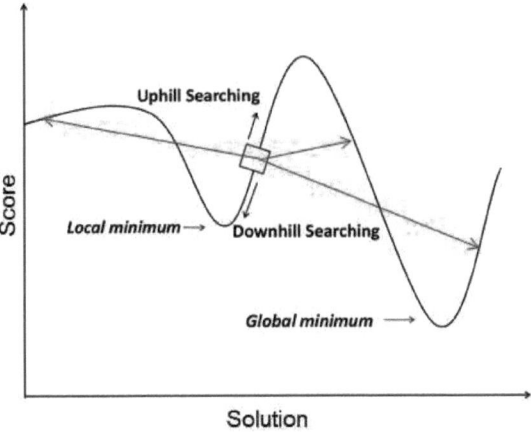

Solution

Figure 3: Simulated annealing is an optimization algorithm that could be implemented with Marxan. The main advantage of this algorithm is that it is able to leave local minimum (red arrows). Hence, by allowing these stochastical bad moves it avoids getting stuck at local minimum.

2.3.1 Input Files

Marxan requires at least four input files to run an optimization. First, a basic input parameter file (input.dat) is needed. It includes general parameters, the number of iterations, the information on the other input files, the input and output path and further settings that has to be done (Appendix 1). Second, there is the species file (spec.dat) that contains each conservation feature, the corresponding targets and species penalty factor (SPF). The target gives the number of individuals of each species that is intended to be in the solution. If this target is not met, the SPF will be summed up with the cost of the solution to a score. Marxan minimizes this score, and, thus, the best solution is the one with the lowest score. Third, the planning unit file (pu.dat) is required. This file is a list of the user-defined planning units, their respective costs and their status. The status determines whether a planning unit is locked in or out

11

of the initial and final reserve system (Tab. 1; Game and Grantham 2008). Finally, there has to be a file that connects the conservation feature information with the information on the planning units. This file is named planning unit versus conservation feature file (puvspr.dat).

Table 1: Status value in the planning unit file

Status	Meaning
0	The PU is not guaranteed to be in the initial (or seed) reserve system, however, it still may be. Its chance of being included in the initial reserve system is determined by the 'starting proportion' specified in the Input Parameter File (input.dat)
1	The PU will be included in the initial reserve system but may or may not be in the final solution.
2	The PU is fixed in the reserve system ("locked in"). It starts in the initial reserve system and cannot be removed.
3	The PU is fixed outside the reserve system ("locked out"). It is not included in the initial reserve system and cannot be added.

2.3.2 Output Files

Marxan produces several outputfiles that illustrate the results in different ways. There is one separate output file for each iteration if the same run is repeated several times. These files contain a list of all planning units in the first column. In the second column, there is either a "0" for not included or a "1" for included in the solution. Among these files, there is the best solution file. It contains the same as the others but for the best solution in the series of the repeat runs. The summary information file gives an overview of the iterations of one run. There the results of each iteration could be compared in consideration of the score, the cost, the number of planning units, the number of missing values (here species) and the penalty costs of the found solution.

2.4 Species Data

2.4.1 GIS-Data of Canton Aargau

The online Geoportal of canton Aargau provided different GIS data of the canton. I used five GIS-layers in the course of the analyses (Tab. 2). First, the amphibian inventory layer contained spatial point data of frog species, toad species, newt species and one salamander species records. Second, the layer "Fledermausquartiere" contains recorded nests of several bat species. Third, the CCIs of canton Aargau are digitized in a polygon layer with additional information on the municipality name, the field name, the habitat type and the size. As the amphibian-inventory, the ornithologic and the reptilian inventory contained point data of recorded species. There were duck and bird species in the ornithologic inventory. The reptilian inventory contained saurian and snake species.

Table 2: Used GIS-layers of canton Aargau

Description	Name	Type	Investi-gation Date	Source
Amphibien-inventar 91/92	AGIS.alg_amphibien92	Vector (point)	1.1.1991 to 1.12.1992	© Aargauisches Geografisches Informations-system (AGIS)
Fledermaus-quartiere	AGIS.alg_fledermausquar	Vector (point)	1973 to 1993	© Aargauisches Geografisches Informations-system (AGIS)
Naturschutzge-biete v.kant.Bed. 1:5'000	AGIS.alg_natschgebkb5t	Vector (polygon)	1.1.1993 to 1.12.1996	© Aargauisches Geografisches Informations-system (AGIS)
Ornitholo-gisches Inventar	AGIS.alg_ornithinv	Vector (point)	1.1.1985 to 1.12.1987	© Aargauisches Geografisches Informations-system (AGIS)
Reptilien-inventar: erfasste Arten	AGIS.alg_reptinvea	Vector (point)	1.1.1987 to 1.1.1990	© Aargauisches Geografisches Informations-system (AGIS)

2.4.2 Archive of CCI

A great deal of the data was out of the CCI archive of Canton Aargau. This archive consisted of non-digitized inventory data taken from surveys of conservation areas ordered by the municipalities. The first step was to find if out whether these inventory data belonged to a CCI and if yes to which one. Some inventory sheets were marked with the CCI number; others were allocated by coordinates, field name or map in comparison with the GIS-layers from canton Aargau or swisstopo (Tab. 2, 3).

Table 3: GIS-layer that contained field names

Description	Name	Type	Investigation Date	Source
Pixelkarte	pk25_komb	Raster	2013	© swisstopo

2.4.3 National Inventory of Dry Meadows and Pastures

Another source of species data, mainly for plant species data, was the national inventory of dry meadows and pastures (Trockenwiesen- und weiden von nationaler Bedeutung BAFU, Data contract 31.10.2013). It contained plant species records in all the CCIs that are in the national inventory as well. The records were derived from the years 1997 and 1998.

2.4.4 Unique Taxonomic Identifier

To add a unique taxonomic identifier (ID) to every plant species, the Landolt-number (Landolt 1977) extended because some inventory sheets of the CCI archive already contained it for each recorded species. All other Landolt-numbers could be found in the online database of the national data and information center of the Swiss flora (infoflora.ch 2013).

14

The taxonomic identifiers from the Fauna Europaea (de Jong 2013) were adopted for all the animal species. They didn't overlap with the plant IDs and could be found online for additive records.

2.4.5 Red List Status

Based on the Swiss red lists for animals and plants (Keller et al. 2010, Monnerat et al. 2007, Moser et al. 2002, BAFU 2013), the red list status was added to each recorded species by means of the scientific name. The red list status of each species tha was given in values from 0 to 4 categories before 1994 were cross-walked to the new system that is compatible with that of the IUCN (Tab. 4). The species that had the status extinct, critically endangered, vulnerable or near threatened were considered as "red list species".

Table 4: Categories of red list status (source: BAFU)

Categories before 1994	Valid Categories	
0	EX	Extinct
0	EW	Extinct in the Wild
1	CR	Critically Endangered
2	EN	Endangered
3	VU	Vulnerable
4a-4d	NT	Near threatened
n	LC	Least Concern
-	DD	Data Deficient
-	NE	Not Evaluated

2.5 Cost Calculation

Only the current costs were calculated for each CCI since there was no information on other types of costs. Current costs were defined as costs for relinquishment of management, compensation money and costs for maintenance. Costs for contracts, planning, acquisition and implementation of new conservation areas were not considered. I based the cost calculation on the report of Ismail et al. (2009). After that, maintenance costs depend on ecosystem type (forest, pasture, meadow, bog), slope and patch size. There are minimum and maximum basic costs per ha and year that depend on ecosystem type. The minimum basic amounts were chosen concerning the optimization problem in Marxan (Tab. 5). For the open, land there was a surcharge per hectare and year depending on the slope of the terrain of the conservation area (Tab. 6). Additionaly, there was a surcharge or a reduction per hectare and year for open land depending on the size of the parcel (Tab. 7). Finally, the cost for maintenance of hedges was calculated by considering the value of the direct payments for hedges in Switzerland per hectare and year (ALN Zurich 2013). The length of hedges per CCI was derived from the GIS layer TLM_Baum_Gebüschreihe_2012 (Tab. 8). I multiplied the total length per CCI with a medium width of four meters (ALN Zurich 2013). The total hedge area per CCI multiplied with the 2500 Swiss francs direct payments per hectare hedges resulted in the total cost for the maintenance of the hedges per CCI. This cost was finally added to the sum of the ecosystem type costs per CCI.

Table 5: Basic costs per 1ha per year

Ecosystem type	Swiss Francs
Forest	190.00
Pasture	827.00
Meadow	1443.00
Bog	1517.00

Table 6: Surcharge per hectare and year depending on the slope

Slope	0-18%	18-35%	35-50%	50-80%	>80%
Pasture	3.50	15.00	28.00	45.00	45.00
Meadow	27.00	114.00	170.50	317.50	1810.50
Bog	30.00	156.50	205.00	457.00*	2296.00*

*only virtual, no such case

16

Table 7: Surcharge or reduction per hectare and year depending on the patch size

Patch Size	<0.5ha	0.5-1.0ha	1.0-2.0ha	2.0-5.0ha	>5.0ha
Pasture	21.00	11.00	0.00	-6.00	-8.00
Meadow	248.50	114.00	0.00	-104.00	-212.00
Bog	282.00	141.00	0.00	-155.00	-255.50

The CCI polygons were transferred to a raster of 25m to 25m size since there were CCIs including different ecosystem types and slopes,. This raster was overlaid with the land cover type layer and the digital elevation model layer (Tab. 8). As a result, each CCI was split in several pixels, which contained the information of ecosystem type and slope. As the land cover type layer does not include the three categories pasture, meadow and bog for open land, the description of the ecosystem type per CCI in the attribute table of the CCI layer defined the category. I calculated parcel size by summing up all pixels of the same ecosystem type. The total cost of each CCI was calculated by multiplying the area with the corresponding basic cost, slope surcharge and size surcharge or reduction (Tab. 5, 6, 7).

Table 8: Used GIS-layers for cost calculation

Description	Name	Type	Year	Source
Primärflächen	geolib.pri25_p_07	Polygon	2007	© swisstopo
Geländedaten: DHM25	geolib.DHM25	Raster	1994	© 1995 Bundesamt für Landestopographie
TLM_Baum_Gebuesch-reihe_2012	geolib.swissTLM3D_1_1_BB	Line	2012	© swisstopo

2.6 Analyses with Marxan

I conducted several Marxan runs with a variety of settings in order to find out how Marxan works with real data of the CCIs. Of course the focus was on the question how many CCIs are required to meet the given targets at minimum cost. On one hand I conducted several runs where targets were set for all species equally ("all species runs") other runs in which only the red list species were targeted ("red list species runs"). In these two groups the parameters "target" in the input file spec.dat and "amount" in the input file puvspr.dat have basically been changed. Because the collected data was often presence data, the numbers of individuals of a species in a CCI wasn't known. So first the amount was set for all species to the same values (1, 2, 5 and 10). Second, random values ranging from 1 to 10 were produced to make the data set more realistic (random.org 2014). So, the amount value for each species varied. The remaining parameters in all the input files stayed the same for every run in this study. Thus, the results of very similar runs allow comparability and may illustrate how the optimization algorithm of Marxan works with the real data of canton Aargau. However, the pragmatically defined and simple targets were used to do this experiment with hypothetical goals. Furthermore, by ascending the targets limits could be found where targets could not be met.

In the following subchapters the four input files are described in detail with regard to the data of this study. The different input files were produced with R-Studio the editor for the statistic software R except the input parameter file.

2.6.1 Input.dat

The applied input parameter files were produced with the included file maker inedit.exe. I mainly used the default Marxan parameters as recommended in the manual (Appendix 1). I only changed the number of repeat runs (NUMREPS) from "1" to "10". Marxan conducted ten runs instead of one run for each parameter setting. Thus, the probability of finding a near optimal solution was higher. In addition, I conducted all runs with simulated annealing by setting the runmode (RUNMODE) to "1".

2.6.2 Spec.dat

The three column names of the species input file need to be "id", "target" and "spf". The column "id" should be filled with unique identifiers of each conservation feature (Tab. 9). There were several possible types of conservation features. After Ardron et al. (2010), a conservation feature was defined as an element that the user would like to occur in the final reserve system. So, I defined it as species that could potentially occur in the final reserve system. Depending on the optimization problem, various targets were added to the species. The SPF was set to "2" in all runs.

2.6.3 Pu.dat

One of the first steps before running Marxan was to define the planning units. Based on the collected data, the CCIs of canton Aargau suggested itself to operate as planning units. So, in this study the 376 CCIs were implemented as planning units whereupon the official IDs were adopted. The cost of each CCI was then calculated as it was described in chapter 2.5. The values were rounded and added to the respective CCI in the planning unit file (Tab. 10). The status as it was described in table 8 was set to "0" for all CCIs in every run. The complete list of the CCIs and their respective cost was added to the appendix 1.

2.6.4 Puvspr.dat

The planning unit versus conservation feature file contained a list with the information of which conservation feature occurred in which planning unit. In the first column the conservation feature IDs were listed as many times as they occur in different planning units. In the second column the planning unit IDs were arranged next to the conservation feature they include (Tab. 11). So each planning unit was listed as many times as they contain conservation features. The information of how many of the respective conservation feature existed in a particular planning unit, was listed in the third and last column "amount". For this study every unique record of all plant and animal species was listed. As mentioned the amount was changed depending on the conducted run.

Table 9: spec.dat

id	target	spf
8	1	2
14	1	2
18	1	2
21	1	2
22	1	2
27	1	2
...	...	...

Table 10: pu.dat

id	cost	status
1	4357	0
2	636	0
3	1942	0
4	1477	0
5	10'102	0
6	10'702	0
...	...	...

Table 11: puvspr.dat

species	pu	amount
8	330	1
8	362	1
14	159	1
18	25	1
18	162	1
21	28	1
...	...	...

2.6.5 Optimization Runs

In table 12 the target and amount of each run was listed. The runs were named as indicator for the settings. So, for example "T1A1" means that the target was set to "1" (T1) and the amount was set to "1" (A1). The target value "1" meant that one individual of each species should occur in the final reserve system. The amount value stood for the number of individuals per CCI of each species. In several runs the random values (R) were set for the amount column. When only the red list species were targeted, "RL" was added to the run name. Assuming the ratio of the target and the fix amount were critical to the best solution different runs with the same ratio were conducted (e.g. T1A1, T2A2, T5A5 with ratio 1). Because the amount is at least as high as the target, I expected similar best solutions for runs T1A2 and T1AR. For these reasons there were eight unique runs. These unique runs were highlighted in bold face in table 12.

20

Table 12: Optimization Runs where the targets are the same for all species; R means random amounts

Run	Target for not threatened species	Target for red list species	Amount
T1A1	1	1	1
T1A2	1	1	2
T2A1	2	2	1
T2A2	2	2	2
T5A1	5	5	1
T5A2	5	5	2
T5A5	5	5	5
T10A1	10	10	1
T10A2	10	10	2
T10A5	10	10	5
T1AR	1	1	R
T2AR	2	2	R
T5AR	5	5	R
T10AR	10	10	R

I conducted another seven runs in which only the red list species were targeted (Tab. 13). Since these species were not found in numerous CCIs the target was only set to "10" when implementing the random amounts. These random amounts were adopted from the runs above and thus stayed the same for all the runs with random amounts.

Table 13: Red list species runs. R stands for random amounts

Run	Target for not threatened species	Target for red list species	Amount
RLT1A1	0	1	1
RLT2A1	0	2	1
RLT5A1	0	5	1
RLT5A2	0	5	2
RLT2AR	0	2	R
RLT5AR	0	5	R
RLT10AR	0	10	R

3. Results

3.1 Specie Data

Merging the data of the different sources resulted in 17'460 records in 339 CCIs. These records minus the double counted due to different record years then turned to 14'329 records. They consisted of 832 plant species and 369 animal species, i.e. 1201 species in total. For 37 CCIs no species data were available. In total there were 1729 records of red list species in 278 CCIs (Tab. 14). These records consisted of 714 plant species records and 1015 animal species records. These records derived from 97 red list plant species and 99 red list animal species.

Table 14: Summary information red list species data

Records	
Number of records of total red list species	1729
• Number of records of red list plant species	714
• Number of records of red list animal species	1015
Species	
Number of total red list species	196
• Number of red list plant species	97
Status CR – critically endangered	2
Status EN – endangered	24
Status VU – vulnerable	71
• Number of red list animal species	99
Status CR – critically endangered	9
Status EN – endangered	27
Status VU – vulnerable	63
Conservation Areas	
Number of CCIs with red list species	277

The CCI with the number 88 had the maximum species number in both cases (Figs. 4, 5; highlighted in light blue). These were in total 360 species including 45 red list species. Thereby it was the CCI with the most recorded species by far. With 241 and 222 species in total and accordingly 23 and 18 red list species, the CCIs with the number 362 and 159 respectively were on the second and third position in the species ranking (in red in Figs. 4 and 5).

In both of the two figures the category with the lowest numbers of species (green) was the one with the most CCIs within. Next to the 37 CCIs that had no species data at all there were 223 CCIs that had 1 to 51 species in total. Regarding the red list species there were another 62 CCIs that had no species records. For 184 CCIs 1 to 6 red list species were recorded.

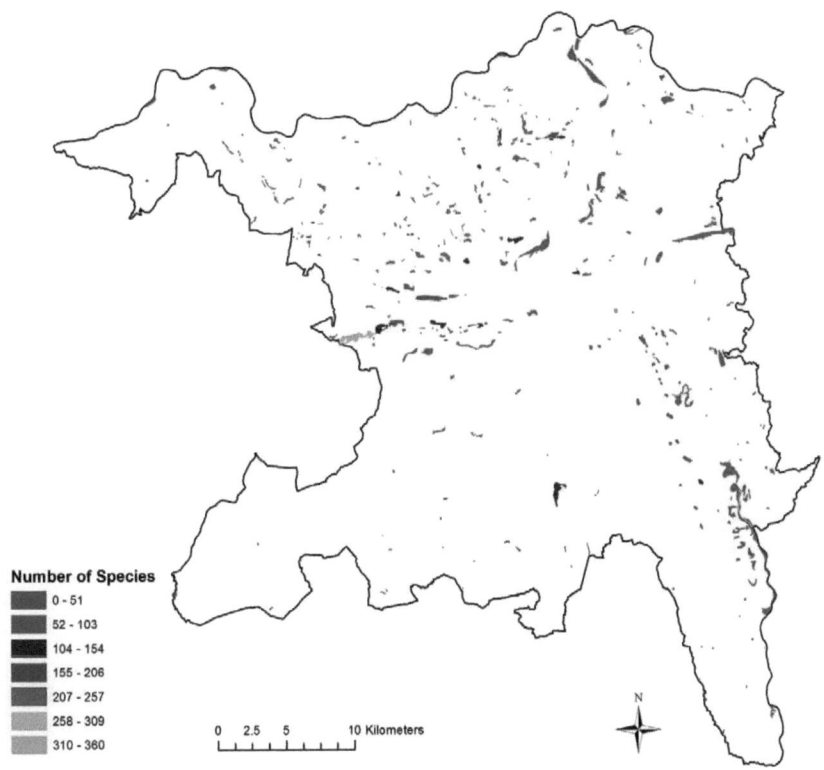

Number of Species
- 0 - 51
- 52 - 103
- 104 - 154
- 155 - 206
- 207 - 257
- 258 - 309
- 310 - 360

0   2.5   5   10 Kilometers

N

Figure 4: The number of species per CCI. The CCI 88 in light blue was the one with the most recorded species. In contrary, there were no data available for 37 CCIs. For the main part, 223 CCIs, 1 to 51 species were recorded.

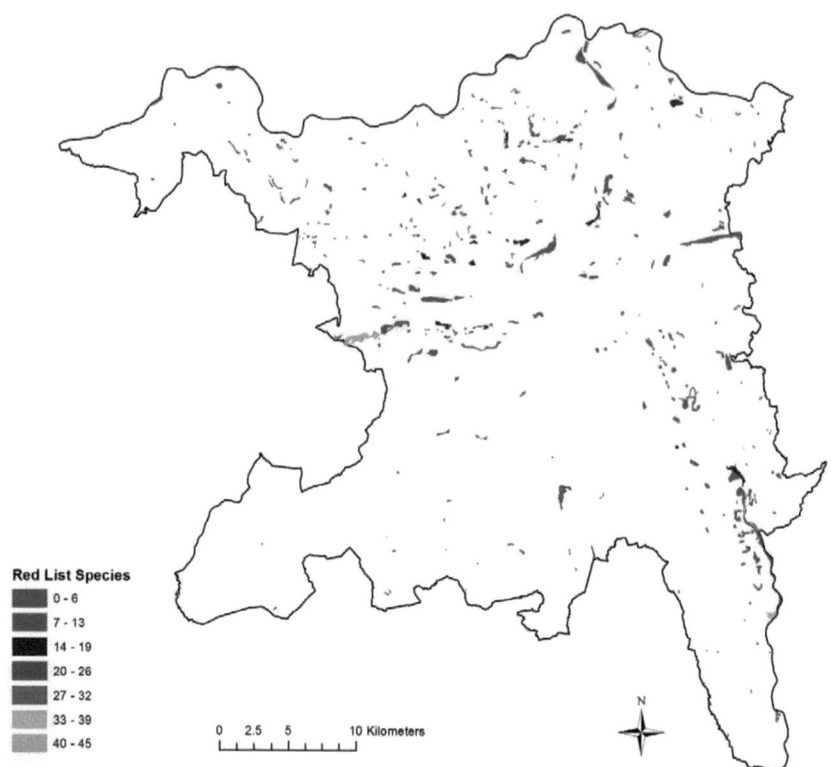

Red List Species
- 0 - 6
- 7 - 13
- 14 - 19
- 20 - 26
- 27 - 32
- 33 - 39
- 40 - 45

0 2.5 5   10 Kilometers

N

Figure 5: The number of red list species per CCI is illustrated in seven equal intervals. Here, even for 99 CCIs no data were available. For another 148 CCIs only 1 to 6 red list species were recorded. Again, the CCI 88 had the maximum number of red list species.

3.1.1 Species with Red List Status Critically Endangered

There were 11 species with red list status CR (critically endangered) in the collected species data (Fig.6-16). Two of them were plant species and nine were animal species. The plant species were both phanerogams (*Sium latifolium, Dactylorhiza maculata*). The animal species consisted of one snake species (*Vipera aspis*), two butterfly species (*Satyrium pruni, Cupido argiades*), three dragonfly species (*Leucorrhinia pectoralis, Leucorrhinia caudalis, Sympetrum*

24

pedemontanum), one mammalian species (*Castor fiber*) and two bird species (*Lanius senator, Vanellus vanellus*).

Figure 6: *Sium latifolium*   **Figure 7:**
Dactylorhiza
maculata

Figure 8: *Vipera aspis*

Figure 9: *Satyrium pruni*

Figure 10: *Cupido argiades*

Figure 11: *Leucorrhinia pectoralis*

Figure 12: *Leucorrhinia caudalis*

Figure 13: *Sympetrum pedemontanum*

Figure 14: *Castor fiber*

26

Figure 15: *Lanius senator*   **Figure 16:** *Vanellus vanellus*

3.2 Calculation of Current Costs

The current cost per year for each CCI added up at maximum 54'620 Swiss francs. This maximum cost was reached by the CCI with the number 60 (highlighted in green in Fig. 17). This CCI is the well-known Boniswiler Ried at the lake of Hallwil. It is the largest remaining low moor in canton Aargau. By far it was the most expensive CCI, so there was no CCI within the next lower category (brown). Thereafter, only one CCI was in the next lower category (dark blue), number 88 the same CCI with the most recorded species. The costs for another two CCIs reached the medium level category (pink). They are located in the south east of the canton at the riverside of Reuss. The categories highlighted in red and orange covered 11 and 35 CCIs respectively. Finally, for 326 CCIs and thus for the main part of the CCIs, the current cost was estimated not to be higher than about 7800 Swiss francs per year (light blue). The CCI with the least cost was number 62 that consisted of a pond. A detailed list of the current cost for each CCI could be found in the appendix 1.

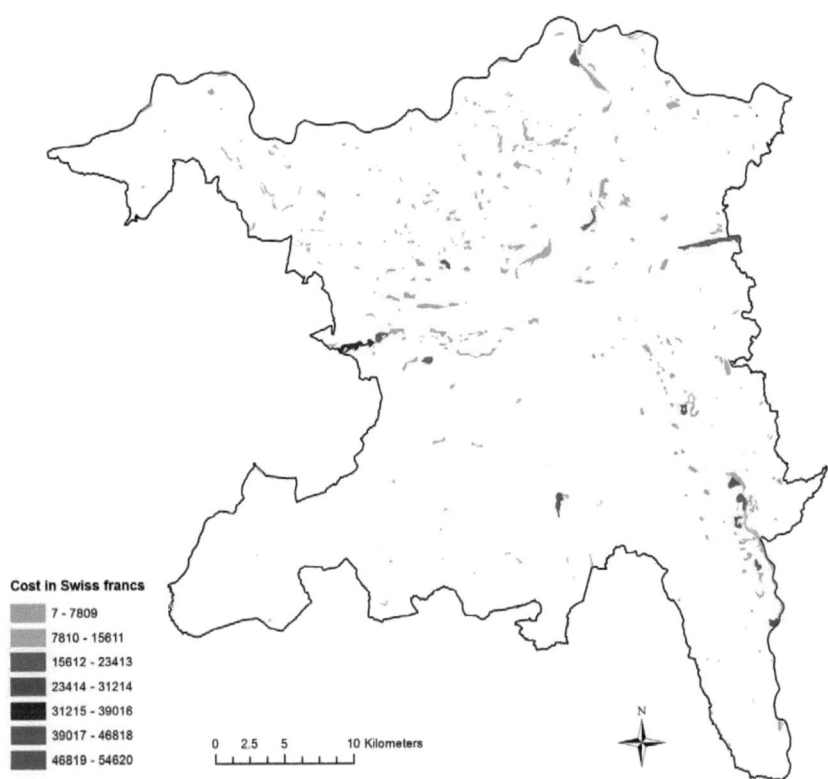

Cost in Swiss francs
- 7 - 7809
- 7810 - 15611
- 15612 - 23413
- 23414 - 31214
- 31215 - 39016
- 39017 - 46818
- 46819 - 54620

0   2.5   5   10 Kilometers

N

Figure 17: The current costs per year for each CCI was calculated and illustrated in seven equal categories. In general, these costs were not higher than 10'000 Swiss francs. There were some exceptions as the most expensive CCI, the Boniswiler Ried at Lake Hallwil (green).

To bring the species data together with the cost data, each dot in figure 18 represented one of the 376 CCIs. In general the CCIs had a total number of species within 0 and 100 at a yearly cost of up to 20'000 Swiss francs. There were only a few outliers. On one hand there were the CCIs with number 88, 362 and 159 which had more than 200 species. On the other hand there were the expensive CCIs as the number 60 and again 88. The figure 18 illustrated that there were a couple of species-rich CCIs that had yearly costs less than 10'000 Swiss francs. Even though there were CCIs that had yearly costs higher than

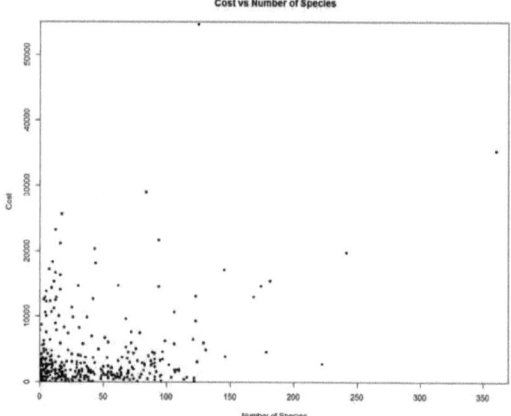

Figure 18: The calculated costs were plotted against the number of recorded species in each CCI. The most CCIs had a number of species from 0 to 100 and yearly current costs in between 0 and 10'000.

10'000 Swiss francs but had not more than 50 recorded species. As the current cost per CCI was calculated per hectare they were expected to correlate with the respective size of a CCI. The plotted data in figure 19 showed this correlation. Thus, larger CCIs had higher total costs.

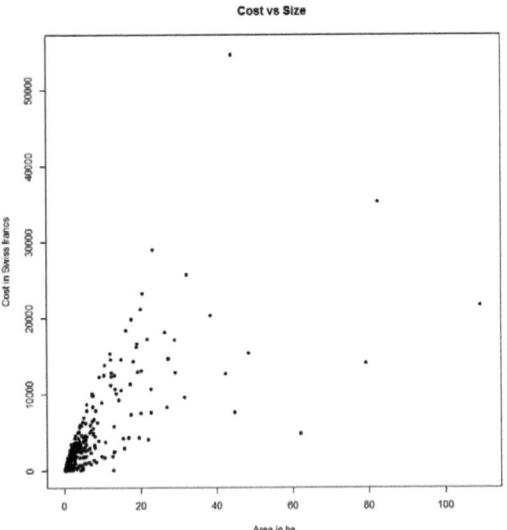

Figure 19: The calculated cost of each CCI was plotted against the respective size in hectares. As the cost was calculated per hectar and then summed up for each CCI, it was expected to correlate with the size.

3.3 Best Solutions of the All Species Runs

Targets could be met in five of the 14 optimization runs (see red dots in Fig. 20). Not all of the species were involved in the best solution of a final reserve system in the remaining nine runs. As expected the best solutions were mostly the same when the ratio of target and amount was the same. Following groups emerged to have the same or nearly the same best solution (Tab. 15).

31

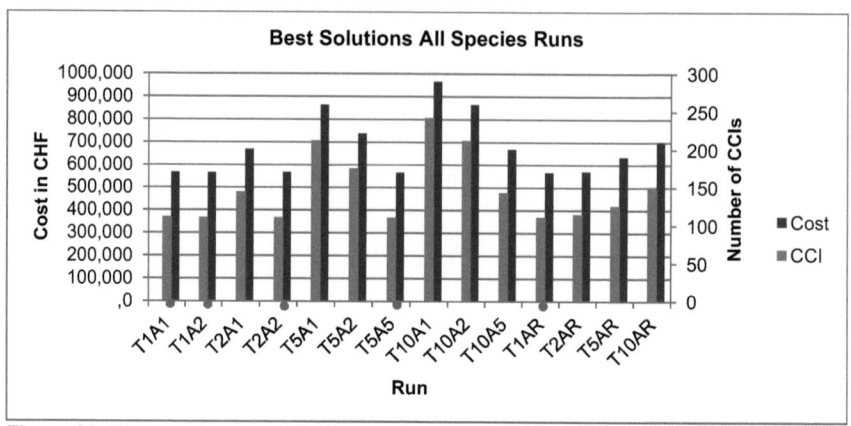

Figure 20: The cost and the number of selected CCIs of the respective best solution of each run was compared. The higher the targets and the lower the amount the higher were the costs and the number of CCIs for a optimal reserve system.

Table 15: Similar runs grouped

Group 1	Group 2	Group 3
T1A1	T2A1	T5A1
T1A2	T10A5	T10A2
T2A2		
T5A5		
T1AR		

In group 1, with target-amount ration of 1 or below, the targets could be met. When only the eight unique ratios were considered, the targets could be met for runs where the amount was at least as high as the targets (e.g. T1A1, T1A2). Targets were not met when the amount was lower than the target (e.g. T2A1, T5A1). If each of these groups (Tab. 15) were seen as one run, targets could be met in only one of the eight unique runs. In group 2, targets were not met for 390 species. In group 3, targets were not met for almost 60% of the species (711). In fact in run T10A1 for even 890 species targets were not met (Tab. 16).

Table 16: Number of missing species in the best solution

Run	Missing Species
Group 1	0
T2AR	29
T5AR	162
Group 2	390
T10AR	426
T5A2	549
Group 3	711
T10A1	890

Overall with the lowest demands (T1) at least 110 of the CCIs were required (group 1). The best solutions of the runs T2AR, T5AR and T10AR involved 115, 127 and 148 CCIs respectively. In the runs of group 2 and 3, 144 CCIs and 212 CCIs were chosen, respectively. For the best solution with the settings of run T5A2 and T10A1 Marxan computed 175 and 242 CCIs respectively. As expected, the best solutions spanned more CCIs the higher the targets and the lower the amounts (Fig. 20).

The respective cost of the best solutions ranged from 564'037 CHF (in T1A2) to 174'163'829 CHF (in T10A1) at which the cost was higher when the targets were not met. In general, the higher the ratio of target to amount, the higher is the risk of not meeting the targets and, thus, the higher the cost of the best solution (e.g. T10A1). This is due to the fact that more CCIs are needed the higher the amount is.

3.4 Best Solutions of the Red List Species Runs

There was one best solution among the seven red list species runs that met all the targets (Fig. 21). It was the run with the lowest target RLT1A1. In the remaining best solutions at least six and maximum 132 species were not involved (Tab. 17). This maximum of missing species was about 65% of the total red list species. In general, the same

33

characteristics as above could be seen in the selection of CCIs for the best solution. Targets were only met when the amount was at least as high as the targets. But since there were less species the best solutions absolutely involved less CCIs. In fact Marxan chose 48 CCIs in the best solution of the run RLT1A1 which had the lowest target. An additional two CCIs were included in the best solution of run RLT2AR. Further, the best solutions of the runs RLT5AR, RLT10AR and RLT2A1 consisted of 58, 71 and 72 CCIs respectively. Finally, the best solution with the most missing species was the one that contained the most CCIs.

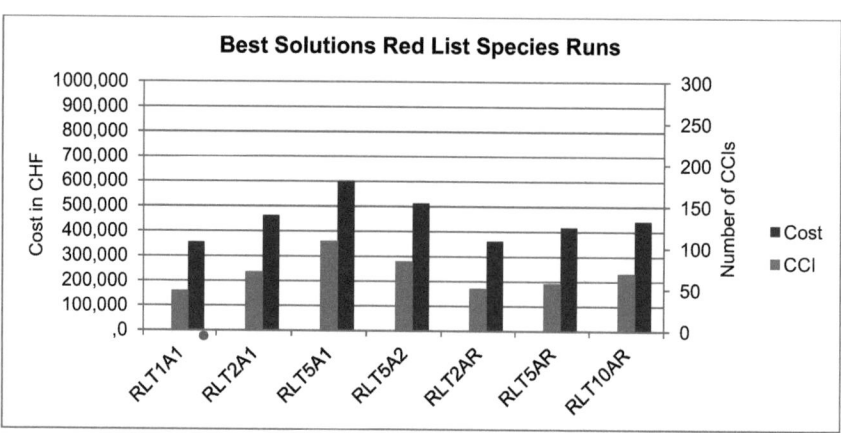

Figure 21: The costs and number of CCIs of the best solutions of the red list species runs were plotted in bars. The number of CCIs in a solution ranged from 44 to 108; the costs from 355'321 to 598'772 Swiss francs.

Table 17: Number of missing species in the best solutions of the red list species runs

Run	Missing Species
RLT1A1	0
RLT2AR	6
RLT5AR	30
RLT2A1	73
RLT10AR	90
RLT5A2	110
RLT5A1	132

3.5 Average Cost per CCI of Best Solutions

In comparison to the best solutions with all the species involved, the cost per CCI was higher in the best solutions with the red list species. Even the lowest average cost in the best solution of run RLT5A1 was higher than the highest average cost of the runs in group 1 (Fig. 22). In both cases the average cost emerged in the best solution where all targets could be met (group 1 and RLT1A1). These two solutions involved the lowest number of CCIs. This indicates that Marxan chose fewer, but more expensive, CCIs to meet the given targets.

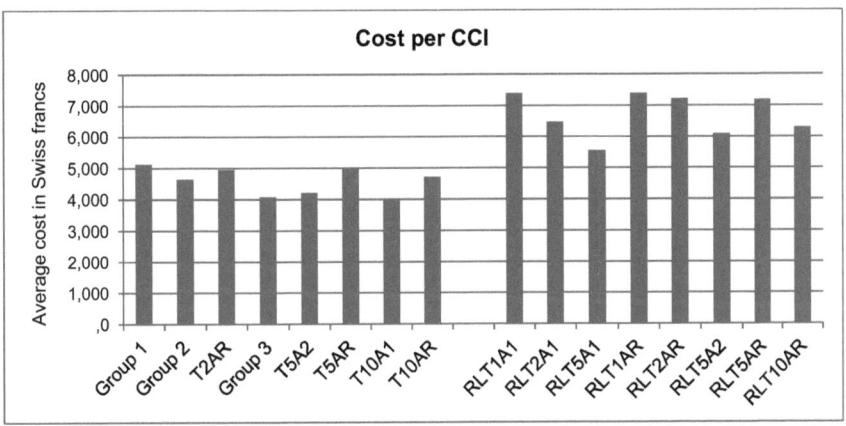

Figure 22: The average cost per CCI in the respective best solution of each run was calculated and compared. The result showed that even the lowest average cost in the best solution of run RLT5A1 was higher than the highest average cost of the runs in group 1.

3.6 Selection Frequencies

As another result, the selection frequency of each CCI over the best solutions of the eight unique all species runs was calculated (Fig. 23). This frequency ranged from 0 meaning never selected in a best solution to 8 meaning selected in all the best solutions. The colors corresponded to the seven equal categories of this frequency and correspond to the colors used in the maps in chapter 3.1 and 3.2. The colors in this map

35

illustrates that a big part of the CCIs was never been selected in these eight runs (green). Contrarily another large part was involved in every best solution (light blue). About 134 CCIs were never selected while 105 CCIs were always included in the best solution (Fig. 24). Seven to 33 CCIs had a selection frequency in between 0 and 8.

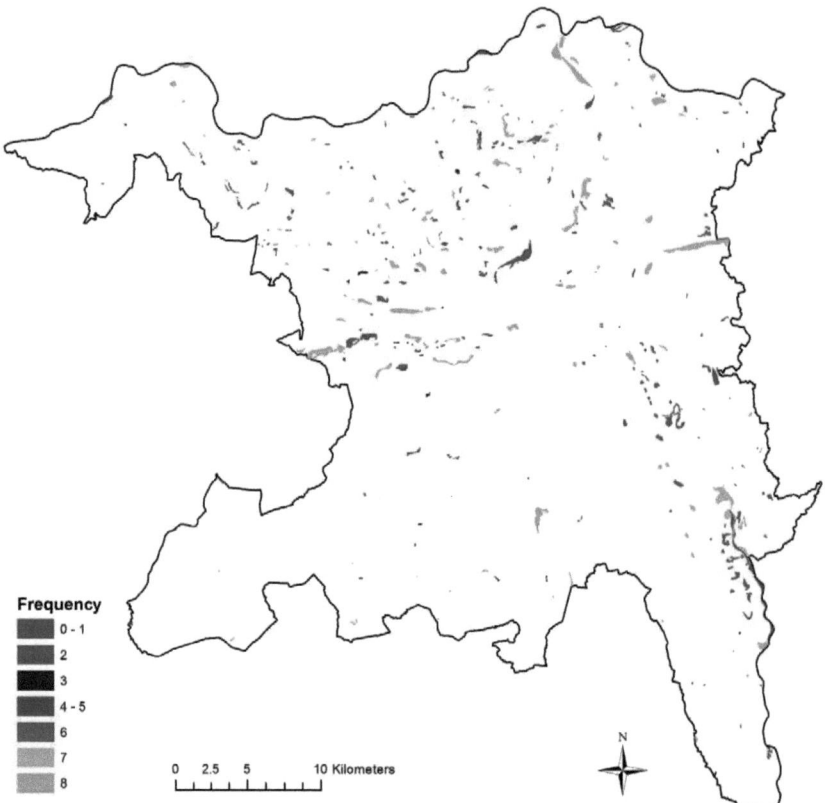

Figure 23: The selection frequency of each CCI over the eight unique all species runs was illustrated. 105 CCIs were included in every best solution. In contrast, another 134 CCIs were never selected.

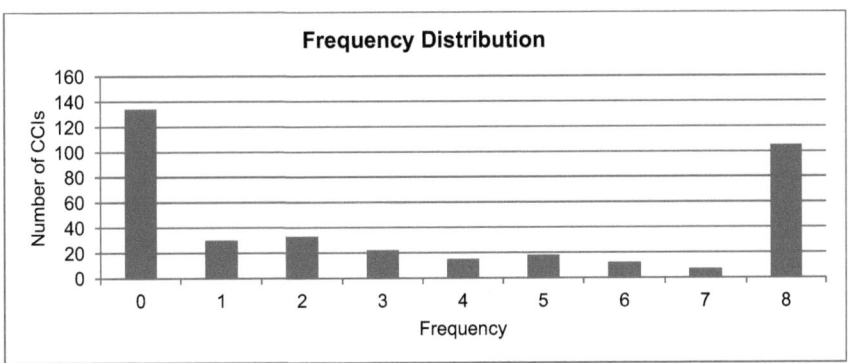

Figure 24: The Number of CCIs was plotted as a bar for each frequency. This frequency distribution illustrated that a CCI was either selected never or in every best solution.

The same as above was done for the red list species runs. A similar result as above emerged (Fig. 25). There were a lot CCIs with a selection frequency either in the lowest (green) or in the highest (light blue) category. But this time the number of CCIs that were never involved in a best solution, was twice as high as with the all species runs (Fig. 26). This time, even 268 CCIs were never selected in a best solution. Contrarily the number of CCIs that were always selected was quite low. Only 44 CCIs were involved in every best solution. Each of these 44 CCIs was in the highest selection frequency category from above too. If the two maps were compared, especially the large CCIs in light blue attracted attention (see red circles in Fig. 25).

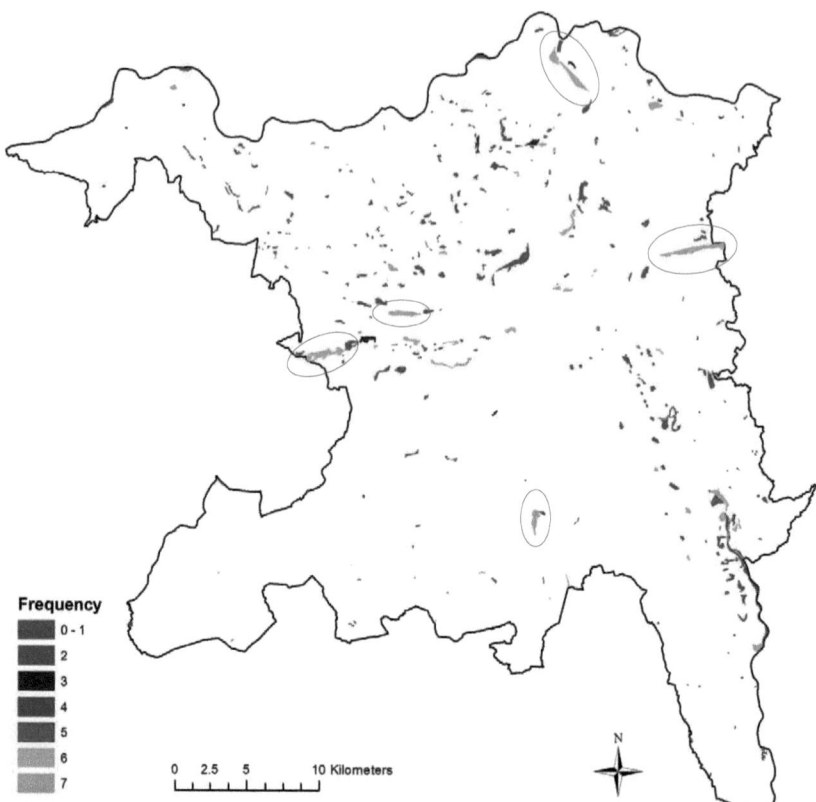

Figure 25: The selection frequency of the seven best solutions of the red list species runs was evaluated. The highest category contained 44 CCIs the lowest 268 CCIs. Some of the largest CCIs were included in every best solution of the all species runs as well as in every best solution of the red list species runs (red circles).

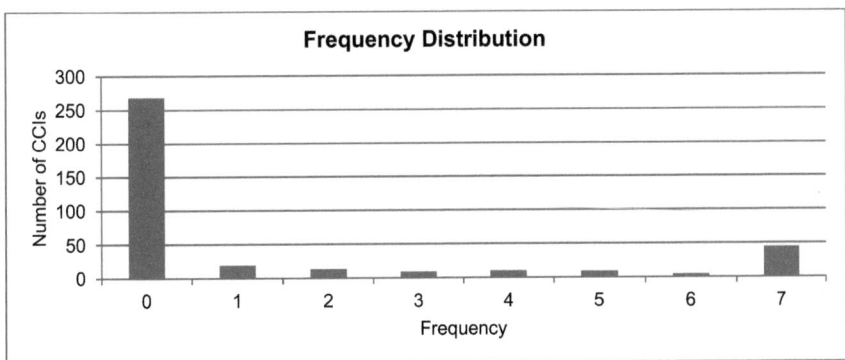

Figure 26: The the number of CCIs per selection frequency of the seven red list species runs was calculated. Here, the distribution was a more extreme version of the frequency distribution of the all species runs. The main part of the CCIs was never included; 44 CCIs were selected in every best solution.

4. Discussion

4.1 Species Data Base

When Marxan runs an optimization it is assumed that the species data are complete. But obviously in this study the data imperfectly covered the species occurrence in all the CCIs. It's about the Linnean shortfall (Whittaker, Araujo et al. 2005) that referred to our lack of information of how many, and what kind, of species there are. That means, we still have incomplete knowledge of the present biodiversity. In addition, I lacked information on the abundance of each species occurred per CCI. Thus, the results of this study should be interpreted carefully. They only illustrate which CCIs Marxan selects with the given information. Even if this analysis with Marxan gave some hints about which CCIs contribute to the protection and preservation of biodiversity, it was not meant to support decisions. This study seeked to apply this systematic-conservation tool as a preliminary analysis. Its results did not intend to be implemented in practice.

In addition to the incompleteness of the species data, another species problem existed. Namely, the inventory data that I used were possibly biased in two ways. First, in this study species data were often based on field inventories of a specific species group, e.g. birds. Second, some inventories biased towards encountering rare or otherwise interesting species. In general, there is no comprehensive species inventory where all occurring species of a conservation area are recorded. Finally, there will never be a guaranty of complete species inventory even if it was sought to.

4.2 Cost Calculation

The estimations of the costs for each planning unit (here CCI) only included the current cost for relinquishment of management, compensation money and costs for maintenance. In fact, there were more costs than just these. For example, there are opportunity costs, implementation costs and even society dependent, non-monetary costs. As mentioned, the cost calculation was done only for the current cost for several reasons. It would be a challenging task to get appropriate estimations for not these additional costs that I did not consider. In addition, Marxan does not need complete cost information. It is up to the user to determine which costs Marxan uses in an optimization. If there were a specific project to plan a new or expanded conservation area network with a cost calculation for a defined time horizon, I recommend inclusions of implementation and opportunity costs. This would result in an absolute cost for the final reserve system because it would be a more comprehensive cost calculation. In this study it was necessary to discriminate the CCIs through a meaningful cost value. Therefore, the current cost per year was considered suitable. In addition, because the cost calculation was conducted per spatial unit and then summed up for each CCI the size of the respective CCI was incorporated. Thus Marxan involved land use as well as monetary costs in the optimization. Furthermore, the applied cost calculation could be adopted as a method for future cost calculations in conservation planning in Switzerland.

4.3 Marxan Analysis Results

The outputs of Marxan indicate that there are some irreplaceable CCIs in canton Aargau. These 44 CCIs evidently contain a major part of the present biodiversity depending on the available data. Even some of the most expensive CCIs were always selected in the best solution. Hence, these CCIs seem to be essential for the reserve network and, thus, meet with the principle irreplaceability. On the contrary, there were a lot of CCIs that were never selected in a best solution. Several reasons could be figured to explain this. First, for 37 CCIs no data was available, thus they didn't contain any conservation features while they had current costs. Marxan didn't select them because it runs an optimization in consideration of benefit (conservation features) and costs (current costs). In these cases, there were only costs without benefit. Second, some CCIs contain species that could be found in another CCI at lower costs. This is more likely the fewer species a CCI contain. Hence, CCIs with few species were never selected if the species were found in another CCI at lower costs.

The analyses with Marxan indicate further that there were several species that occur in one single CCI. For this reason, the target having each species at least twice in the final reserve could not be met when the amount was set to "1". The same target was not met even when the amount was randomly set. This implies that the distribution of some species is poor in this reserve system. Obviously, the present database contributes to this conclusion.

The higher average cost per CCI of the best solutions of the red list species runs was an interesting result too. It could be casually determined as follows, if this was not an accidental result. We assume that red list species are threatened because they make high demands on their habitat. Further, it is expensive to provide these habitats or the opportunity costs are very high. Thus, only a few adequate places for these species have remained and for which reason some species have got threatened. I recommend conducting another study to proof or disprove this thesis.

4.4 User-defined Targets

A key question in systematic conservation planning is, how many individuals of each species are required in the reserve system to ensure their persistence? This knowledge is essential when minimizing the loss of biodiversity is the objective of conservation planning. But as in this study, this knowledge is usually lacking. It would have been used to set the targets for each species in the Marxan analyses. For this reason, a set of targets was implemented in these Marxan analyses. In this way, it could be dealt with the incomplete information because the selection frequencies gave a hint of the importances of each CCI.

The arisen problem is not new for conservationists and there were several disputed attempts to formulate rules of thumb (Hanski, Moilanen et al. 1996). After Moilanen, Wilson et al. (2009) there are two approaches that account for this problem but these have limitations too. First, there are adaptable metapopulation models which operate with species-specific parameters. These models are based on the metapopulation theory (see Hanski and Gaggiotti 2004). Second, there are different types of spatial population viability analyses that integrate information on life history of a species.

There is no way to get around such modelling approaches in the field of systematic conservation planning. But it is a difficult task to decide for an appropriate method for a current problem although the methods are available. At this point, further research has to be done in Switzerland before applying systematic conservation planning in practice.

4.5 Number of Species versus Species Composition

In this study, biodiversity was only described by the number of species at a site. But as the principle of comprehensiveness (Moilanen, Wilson et al. 2009) indicates there should be consideration given to species composition and genetic diversity, to habitat types and ecological processes. These are defined as components of biodiversity as well. While species composition and genetic diversity could be technically handled with Marxan the available information is not sufficient in this case. It is possible to run Marxan by defining the different habitat types as conservation features. This could be done in a supplementary analysis to focus on the structural diversity of conservation areas in canton Aargau. It is likely that information on ecologically important structures will remain incomplete as well. But there are several GIS-layers that would provide additional, detailed information on these additional structures. However, Marxan doesn't model ecological processes (Ardron et al. 2010). Thus, other methods have to be applied in order to address the principle of comprehensiveness in the consideration of ecological function. But it is more likely that there is a lack of appropriate data than a lack of appropriate methods.

4.6 Biodiversity Processes

One ecolgical process that influences genetic diversity is migration. To ensure migration of species in between conservation areas, connectivity is crucial. Since the input file for Marxan didn't offer any information of the spatial location of the CCIs, the present analysis makes no statement about the connectivity of the chosen CCIs. Thus, the effectiveness of a final reserve system concerning migration and genetic exchanges was not evaluated in this study. This problem could be solved in one of two ways. First, one could introduce a connectivity matrix to Marxan. The connectivity matrix contains numbers that account for the connectivity between planning units. If one planning unit is chosen for the solution, and the other is not, the values influence costs. Second, the whole study area of canton Aargau could be split up in a grid of regular planning units. These planning units then would share a part of their boundary with surrounding planning units. The information on the shared boundaries could be implemented in Marxan. With a boundary length modifier option one can determine how compact and thus connected the overall reserve system should be. However, these two approaches could only be applied if the database is sufficient. In this study, the data was not adequate for the connectivity matrix, nor for implementation of regular grid planning units. For these reasons the CCIs were defined as planning units even though there was this trade-off.

4.7 Dynamics

The input data for an analysis with Marxan are always snapshots in time and space. Hence the output here merely shows a near optimal solution for the moment, but not necessarily for the future. The principle of persistence is not involved directly in this analysis. But there have been studies that seek to overcome this problem. Pressey et al. (2007) dealt with the snapshot problem too. They concluded that conservation planning is not effective in providing for conservation of biodiversity processes and neither are dynamic threats considered. As an approach to include dynamics Pressey (2007) focus on modelling future conditions. This approach was implemented by Loyola et al. (2013) when they applied models to generate data about potential future ecological niches and climate. They then estimated future species occurrences based on the combination of the output of these models. Finally, they ran Marxan twice, once for the current species occurrence data and again for the future species occurrences, and subsequently compared the outputs. With this method they addressed to the problem of the static character of species occurrence data. If one projection of species occurrence was adequate is another question. Since all models make some assumptions, their output should be critically interpreted.

4.8 Transfer Gap between Science and Practice

After all the mentioned points the gap between science and practice in conservation planning in Switzerland has to be closed. Since systematic conservation planning needs a comprehensive planning process, it would take a long time until satisfactory conservation plans are on hand and could be implemented. In fact, there are concepts on how to implement systematic conservation planning in reality (see chapter 2.2 and Groves, Jensen et al. 2002). But prior to going through these steps, the idea of systematic conservation planning has to be made available for practicioners. This could be done by conducting workshops, a planning guidance or even an information center.

5. Conclusions and Future Prospectives

This study was a first attempt to run a systematic conservation-planning tool with real data on plant and animal species occurrence in a region in Switzerland. The results illustrated that Marxan could identify some CCIs that seems to be irreplaceable for the reserve network in canton Aargau. Further, the outputs of Marxan were constrained by the availability of data. Therefore, these outputs should not be used as decision support. The calculation of the current costs was founded and transparent. Due to its flexibility it could be used as a method in further studies in Switzerland. In general, this study showed a rather technical approach that needs a lot of more work to be done.

In my opinion, the concept of systematic conservation planning should be andvanced in Switzerland. It is a comprehensive approach as it seeks to consider all relevant factors. Further, Marxan analyses should be used as a decision support only. However, I recommend improving several things. First, the database of the conservation features should be more appropriate. Second, further studies should be conducted to find sufficient methods that address the problems of incomplete database, setting specific targets and including dynamics. I think, there already exist some models that could be a possibility to deal with these problems. But, the selection of such a method should be reflected for every single analysis, because each problem in conservation planning is different. Finally, there is another version of Marxan that works with spatial zonation. I suppose that Marxan with zonation has a high potential to operate with more complex problems. Thus, this software should be applied for Switzerland too.

I believe that systematic conservation planning would be best practice for Switzerland too. Therefore, it is important to close the gap between science and practice. The concept of systematic conservation planning should be made available for the Swiss conservationists in practice. This information exchange could be advanced by conducting workshops or seminars in conservation planning.

Acknowledgment

First I want to thank my supervisors Peter B. Pearman and Felix Kienast for supporting my master thesis. They organized a work place at WSL, answered to current questions and helped me clarifying some communication problems. Thanks go to Dirk Schmatz and Melanie Germann who helped in the digitalization of the collected archive data. Finally, I am thankful that Canton Aargau (Departement of Landscape and Water) provided access to the non-digitized species inventories of conservation areas of cantonal importance.

Tables

Figures

References

ALN Amt für Landschaft und Natur des Kantons Zürich, Fachstelle Naturschutz (2013). Merkblatt Hecken

Ardron, J.A., Possingham, H.P., and Klein, C.J. (eds). (2010). Marxan Good Practices Handbook, Version 2. Pacific Marine Analysis and Research Association, Victoria, BC, Canada. 165 pages.

Bolliger, J., Edwards, T. C., Eggenberg, S., Ismail, S., Seidl, I., & Kienast, F. (2011). Balancing Forest-Regeneration Probabilities and Maintenance Costs in Dry Grasslands of High Conservation Priority. [Article]. *Conservation Biology, 25*(3), 567-576.

Ferrier, S. (2002). Mapping spatial pattern in biodiversity for regional conservation planning: Where to from here? [Article; Proceedings Paper]. *Systematic Biology, 51*(2), 331-363.

Game, E. T. and H. S. Grantham. (2008). Marxan User Manual: For Marxan version 1.8.10. University of Queensland, St. Lucia, Queensland, Australia, and Pacific Marine Analysis and Research Association, Vancouver, British Columbia, Canada.

Groves, C. R., Jensen, D. B., Valutis, L. L., Redford, K. H., Shaffer, M. L., Scott, J. M., ... Anderson, M. G. (2002). Planning for biodiversity conservation: Putting conservation science into practice. [Article]. *Bioscience, 52*(6), 499-512.

Hanski, I. and Gaggiotti, O. E. eds. (2004). Ecology, Genetics and Evolution of Metapopulations.

Hanski, I., Moilanen, A., & Gyllenberg, M. (1996). Minimum viable metapopulation size. [Article]. *American Naturalist, 147*(4), 527-541.

Ismail, S.; Schwab, F.; Tester, U.; Kienast, F.; Martinoli, D.; Seidl, I., (2009): Kosten eines gesetzeskonformen Schutzes der Biotope von nationaler Bedeutung. Technischer Bericht. Available from World Wide Web <http://www.wsl.ch/publikationen/pdf/9625.pdf>. Birmensdorf, Eidg. Forschungsanstalt für Wald, Schnee und Landschaft WSL; Basel, Pro Natura; Bern, Forum Biodiversität, SCNAT. 122 S.

Keller V., Gerber A., Schmid H., Volet B., Zbinden N. (2010): Rote Liste Brutvögel. Gefährdete Arten der Schweiz, Stand 2010. Bundesamt für Umwelt, Bern, und Schweizerische Vogelwarte, Sempach. Umwelt-Vollzug Nr. 1019. 53 S.

Landolt E., Bäumler B., Erhardt A., Hegg O., Klötzli F., Lämmler W., Nobis M., Rudmann-Maurer K., Schweingruber F.H., Theurillat J.P., Urmi E., Vust M., Wohlgemut T. (1977). Flora indicativa: Ökologische Zeigerwerte und biologische Kennzeichen zur Flora der Schweiz und der Alpen; Hautp-Verlag, 768 S.

Loyola, R. D., Lemes, P., Nabout, J. C., Trindade, J., Sagnori, M. D., Dobrovolski, R., & Diniz, J. A. F. (2013). A straightforward conceptual approach for evaluating spatial conservation priorities under climate change. [Article]. *Biodiversity and Conservation, 22*(2), 483-495.

Margules, C. R., & Pressey, R. L. (2000). Systematic conservation planning. [Review]. *Nature, 405*(6783), 243-253.

McDonnell, M. D., Possingham, H. P., Ball, I. R., & Cousins, E. A. (2002). Mathematical methods for spatially cohesive reserve design. [Article]. *Environmental Modeling & Assessment, 7*(2), 107-114.

Moilanen, A., Wilson, K. A., & Possingham, H. P. (2009). *Spatial conservation prioritization: quantitative methods and computational tools*: Oxford University Press New York.

Monnerat C., Thorens P., Walter T., Gonseth Y. (2007): Rote Liste der Heuschrecken der Schweiz. Bundesamt für Umwelt, Bern, und Schweizer Zentrum für die Kartographie der Fauna, Neuenburg. Umwelt-Vollzug 0719: 62 S.

Moser, D., A. Gygax, B. Bäumler, N. Wyler & R. Palese (2002): *Rote Liste der gefährdeten Farn- und Blütenpflanzen der Schweiz*. Hrsg. Bundesamt für Umwelt, Wald und Landschaft, Bern; Zentrum des Datenverbundnetzes der Schweizer Flora, Chambésy; Conservatoire et Jardin botaniques de la Ville de Genève, Chambésy. BUWAL-Reihe «Vollzug Umwelt». 118 S.

Pearce, J. L., Kirk, D. A., Lane, C. P., Mahr, M. H., Walmsley, J., Casey, D., ... Jones, K. (2008). Prioritizing avian conservation areas for the yellowstone to Yukon Region of North America. [Article]. *Biological Conservation, 141*(4), 908-924.

Pressey, R. L., Cabeza, M., Watts, M. E., Cowling, R. M., & Wilson, K. A. (2007). Conservation planning in a changing world. [Review]. *Trends in Ecology & Evolution, 22*(11), 583-592.

Sarkar, S., Pressey, R. L., Faith, D. P., Margules, C. R., Fuller, T., Stoms, D. M., ... Andelman, S. (2006). Biodiversity conservation-planning tools: Present status and challenges for the future *Annual Review of Environment and Resources* (Vol. 31, pp. 123-159). Palo Alto: Annual Reviews.

Whittaker, R. J., Araujo, M. B., Paul, J., Ladle, R. J., Watson, J. E. M., & Willis, K. J. (2005). Conservation Biogeography: assessment and prospect. [Review]. *Diversity and Distributions, 11*(1), 3-23.

Inernet Sources

Bundesamt für Umwelt (BAFU):
http://www.bafu.admin.ch/tiere/07964/08223/index.html?lang=de (14.11.2013)

Canton Aargau:
https://www.ag.ch/de/bvu/umwelt_natur_landschaft/naturschutz/nachhaltigkeit
_2/schuschutzgeb/sschutzgebiet_1.jsp (10.12.2013)

https://www.ag.ch/de/bvu/wald/wald.jsp (10.12.2013)

https://www.ag.ch/media/kanton_aargau/bvu/dokumente_2/raumentwicklung/g
rundlagen_6/raumbeobachtung_1/Bericht_Bodennutzung.pdf (18.2.2014)

Fauna Europaea:
de Jong, Y.S.D.M. (ed.) (2013) Fauna Europaea version 2.6. Web Service
available online at http://www.faunaeur.org (26.11.2013)

Info Flora:
www.infoflora.ch (08.10.2013)

Random Amounts:
www.random.org (06.02.2014)

Figures

Figure 1:
http://upload.wikimedia.org/wikipedia/commons/9/9b/Karte_Kanton_Aargau_2
010.png (18.2.2014)

Figure 3:
http://www.intechopen.com/books/simulated-annealing-single-and-multiple-
objective-problems/simulated-annealing-for-fast-motion-estimation-algorithm-
in-h-264-avc (04.03.2014)

Figure 6:
http://www.infoflora.ch/de/flora/1539-sium-latifolium.html (11.02.2014)

Figure 7:
http://www.fourlangwebprogram.com/fourlang/nl/f_Dactylorhiza_maculata.html
(11.02.2014)

Figure 8:
http://www.artenschutz.ch/cr1.htm (11.02.2014)

Figure 9:
  http://www.vogelwarte.ch/woodchat-shrike.html (11.02.2014)

Figure 10:
  http://www.vogelwarte.ch/kiebitz.html (11.02.2014)

Figure 11:
  http://www.lepiforum.de/lepiwiki.pl?Satyrium_Pruni (11.02.2014)

Figure 12:
  http://www.lepiforum.de/lepiwiki.pl?Cupido_Argiades (11.02.2014)

Figure 13:
  http://www.artenschutz.ch/cr3.htm (11.02.2014)

Figure 14:
  http://www.artenschutz.ch/cr3.htm (11.02.2014)

Figure 15:
  http://www.iucn.org/knowledge/news/?3913/Dragonflies-go-thirsty-in-the-Mediterranean (11.02.2014)

Figure 16:
  http://www.pronatura-aargau.ch/cms/index.php?id=257 (11.02.2014)

Appendix 1

General Parameters	
VERSION	0.1
BLM	0
PROP	0
RANDSEED	-1
BESTSCORE	-1
NUMREPS	10
Annealing Parameters	
NUMITNS	1000000
STARTTEMP	-1
COOLFAC	6
NUMTEMP	10000
Cost Threshold	
COSTTRESH	0
THRESHPEN1	14
THRESHPEN2	1
Input Files	
INPUTDIR	input
SPECNAME	spec.dat
PUNAME	pu.dat
PUVSPRNAME	puvspr.dat
Save Files	
SCENNAME	output
SAVERUN	2
SAVEBEST	2
SAVESUMMARY	2
SAVESCEN	2
SAVETARGMET	2
SAVESUMSOLN	2
SAVELOG	2
SAVESNAPSTEPS	0
SAVESNAPCHANGES	0
SAVESNAPFREQUENCY	0
OUTPUTDIR	output
Program control	
RUNMODE	1
MISSLEVEL	1
ITIMPTYPE	0
HEURTYPE	-1
CLUMPTYPE	0
VERBOSITY	3
SAVESOLUTIONSMATRIX	3

CCI	Spec	Cost	CCI	Spec	Cost	CCI	Spec	Cost	CCI	Spec	Cost	CCI	Spec	Cost
1	74	4'357	53	19	813	106	114	438	159	222	2'757	212	1	1'317
2	2	636	54	56	1'497	107	53	1'132	160	129	6'000	213	26	764
3	74	1'942	55	9	14'350	108	19	2'477	161	12	1'876	214	48	453
4	18	1'477	56	22	4'842	109	0	1'622	162	15	1'792	215	15	10'045
5	5	10'102	57	58	969	110	42	60	163	31	1'498	216	34	5'798
6	9	10'702	58	30	1'266	111	0	1'380	164	1	768	217	76	5'085
7	2	3'724	59	95	854	112	1	1'861	165	49	923	218	37	835
8	6	1'618	60	125	54'620	113	108	1'842	166	50	2'343	219	13	399
9	1	3'477	61	35	2'939	114	2	5'557	167	0	760	220	0	936
10	11	15'333	62	8	7	115	0	2'636	168	22	7'392	221	72	5'985
11	3	5'005	63	17	1'250	116	0	2'386	169	0	515	222	31	266
12	0	1'493	64	44	18'125	117	2	345	170	1	657	223	9	3'962
13	12	23'224	65	15	3'853	118	61	1'292	171	5	12'318	224	9	2'531
14	11	11'194	66	2	2'417	119	85	482	172	8	12'383	225	0	3'093
15	5	4'838	67	62	3'268	120	60	615	173	3	12'558	226	16	2'419
16	131	4'877	68	26	155	121	53	3'710	174	4	10'591	227	6	2'623
17	40	1'821	69	58	1'131	122	83	2'817	175	3	5'713	228	0	963
18	0	559	70	106	5'822	123	6	2'520	176	3	1'940	229	1	964
19	79	761	71	20	660	124	0	711	177	8	6'412	230	1	1'835
20	54	1'468	72	51	6'700	125	91	2'931	178	5	13'823	231	20	42
21	38	2'115	73	71	375	126	30	14'626	179	10	18'388	232	32	35
22	0	468	74	54	6'016	127	5	2'795	180	5	3'574	233	96	174
23	0	129	75	79	7'530	128	71	2'996	181	91	2'319	234	0	686
24	3	1'745	76	74	479	129	122	158	182	63	630	235	1	240
25	30	4'084	77	56	1'217	130	89	2'774	183	70	4'554	236	0	3'377
26	15	728	78	5	1'154	131	82	1'782	184	8	3'781	237	91	4'354
27	1	1'605	79	50	1'366	132	62	14'717	185	12	2'992	238	10	524
28	29	2'653	80	54	697	133	1	927	186	81	2'904	239	12	1'644
29	9	1'582	81	89	4'494	134	5	2'146	187	77	3'233	240	1	66
30	1	1'470	82	28	677	135	95	3'320	188	124	3'094	241	13	12'869
31	4	12'892	83	6	4'282	136	16	1'946	189	43	2'928	242	7	2'631
32	12	1'210	84	1	1'189	137	64	2'102	190	84	29'024	243	28	1'547
33	0	2'671	85	0	367	138	92	1'573	191	12	1'277	244	68	5'264
34	0	592	86	8	3'222	139	73	1'011	192	99	6'288	245	7	17'240
35	35	429	87	62	746	140	4	1'932	193	15	866	246	0	1'162
36	35	539	88	360	35'307	141	19	8'346	194	0	7'863	247	5	182
37	35	915	89	90	4'377	142	7	1'426	195	42	3'706	248	12	16'672
38	0	854	90	122	612	143	0	299	196	3	772	249	2	1'724
39	146	3'901	91	87	1'342	144	12	69	197	0	357	250	9	3'263
40	88	2'585	92	13	447	145	26	9'864	198	5	2'365	251	17	25'661
41	56	2'193	93	43	1'159	146	25	11'403	199	9	2'081	252	12	12'543
42	39	1'600	94	97	4'636	147	10	2'828	200	0	4'263	253	16	21'197
43	35	499	95	12	8'914	148	13	7'904	201	3	4'531	254	1	1'022
44	123	13'135	96	16	5'905	149	6	1'408	202	1	2'784	255	178	4'592
45	83	2'520	97	6	729	150	174	14'674	203	8	2'610	256	16	1'552
46	48	612	98	16	16'282	151	145	17'115	204	9	4'774	257	31	8'253
47	17	239	99	36	976	152	91	378	205	20	314	258	4	754
48	58	393	100	4	1'382	153	17	838	206	1	1'458	259	62	792
49	54	568	101	17	1'007	154	5	557	207	17	2'942	260	46	4'979
50	29	2'996	102	38	1'850	155	20	1'057	208	48	1'934	261	11	1'770
51	25	317	103	91	3'428	156	86	4'013	209	36	3'802	262	1	3'166
52	69	3'645	104	123	9'286	157	24	1'803	210	59	934	263	14	1'231
			105	96	997	158	43	20'328	211	1	683	264	66	312

265	5	43	320	73	2'262	375	67	609
266	39	55	321	57	937	400	68	9'599
267	94	14'577	322	69	756			
268	3	1'893	323	106	10'707			
269	0	118	324	16	14'114			
270	41	6'984	325	1	468			
271	33	1'920	326	65	3'742			
272	4	3'466	327	1	8'689			
273	33	1'650	328	32	2'893			
274	0	706	329	2	69			
275	0	800	330	94	21'689			
276	6	1'311	331	20	297			
277	2	468	332	32	745			
278	22	459	333	63	1'783			
279	0	318	334	40	1'386			
280	23	25	335	0	83			
281	0	2'624	336	1	307			
282	3	3'499	337	1	675			
283	0	1'680	338	6	2'514			
284	38	625	339	2	214			
285	116	707	340	0	303			
286	42	1'508	341	38	234			
287	4	3'346	342	104	709			
288	0	447	343	91	800			
289	20	1'694	344	16	695			
290	181	15'427	345	1	4'220			
291	29	897	346	51	1'012			
292	49	1'703	347	110	1'744			
293	3	129	348	110	1'940			
294	37	2'295	349	38	1'581			
295	70	1'406	350	61	777			
296	38	292	351	69	1'223			
297	64	1'387	352	57	2'206			
298	14	601	353	31	1'702			
299	3	174	354	70	1'694			
300	17	2'162	355	10	2'454			
301	64	951	356	8	950			
302	3	2'621	357	0	1'151			
303	1	1'395	358	0	181			
304	14	1'226	359	32	3'994			
305	42	12'691	360	29	2'153			
306	18	1'191	361	1	379			
307	25	4'212	362	241	19'829			
308	5	7'604	363	102	2'636			
309	72	7'624	364	106	1'868			
310	3	6'257	365	54	570			
311	80	1'221	366	64	2'154			
312	12	508	367	25	3'060			
313	97	3'466	368	88	3'268			
314	121	6'486	369	21	3'068			
315	107	1'600	370	37	9'865			
316	54	2'378	371	0	800			
317	168	12'935	372	70	1'614			
318	77	83	373	32	1'020			
319	41	75	374	64	1'086			

Appendix 3

Plant Species	ID	Red List Status
Abies alba	91	LC
Acer campestre	1988	LC
Acer opalus	1989	LC
Acer platanoides	1986	LC
Acer pseudoplatanus	1985	LC
Aceras anthropophorum	749	VU
Achillea millefolium	3176	LC
Achillea ptarmica	3181	NT
Achnatherum calamagrostis	186	LC
Acinos arvensis	2541	LC
Aconitum compactum	1106	LC
Aconitum lycoctonum	1099	LC
Acorus calamus	561	VU
Actaea spicata	1119	LC
Adenostyles alliariae	3037	LC
Adenostyles glabra	3036	LC
Adoxa moschatellina	2837	LC
Aegopodium podagraria	2241	LC
Agrimonia eupatoria	1560	LC
Agrostis capillaris	190	LC
Agrostis stolonifera	191	LC
Ajuga genevensis	2463	LC
Ajuga reptans	2462	LC
Alchemilla vulgaris	1552	LC
Alisma lanceolatum	151	VU
Alisma plantago-aquatica	150	LC
Alliaria petiolata	1356	LC
Allium carinatum	679	NT
Allium lusitanicum	675	LC
Allium oleraceum	677	LC
Allium scorodoprasum	664	VU
Allium sphaerocephalon	667	LC
Allium ursinum	662	LC
Allium vineale	666	LC
Alnus glutinosa	830	LC
Alnus incana	831	LC
Alnus viridis	828	LC
Alopecurus aequalis	216	VU
Alopecurus geniculatus	215	VU
Althaea hirsuta	2014	EN
Alyssum montanum	1315	VU
Amelanchier ovalis	1655	LC
Anacamptis pyramidalis	755	VU
Anagallis minima	2340	EN
Anemone nemorosa	1137	LC
Anemone ranunculoides	1136	LC
Angelica sylvestris	2204	LC
Anthemis tinctoria	3166	NT
Anthericum liliago	631	LC
Anthericum ramosum	632	LC
Anthoxanthum odoratum	180	LC

Anthriscus nitida	2192	LC
Anthriscus sylvestris	2191	LC
Anthyllis carpatica	1775	LC
Anthyllis vulneraria	1773	LC
Anthyllis vulneraria	1733	LC
Aquilegia atrata	1114	LC
Aquilegia vulgaris	1113	LC
Arabidopsis thaliana	1405	LC
Arabis hirsuta	1416	LC
Arabis turrita	1407	LC
Arctium lappa	2958	LC
Arenaria serpyllifolia	1039	LC
Arrhenatherum elatius	240	LC
Artemisia absinthium	3222	LC
Arum maculatum	563	LC
Aruncus dioicus	1535	LC
Asarum europaeum	864	LC
Asparagus officinalis	636	LC
Asperula cynanchica	2775	LC
Asplenium fontanum	47	NT
Asplenium trichomanes	42	LC
Asplenium viride	43	LC
Aster amellus	3154	LC
Aster bellidiastrum	3162	LC
Athamanta cretensis	2181	LC
Athyrium filix-femina	30	LC
Atriplex patula	934	LC
Atropa bella-donna	2578	LC
Barbarea vulgaris	1389	LC
Bellis perennis	3163	LC
Berberis vulgaris	1213	LC
Berula erecta	2243	LC
Betula pendula	824	LC
Bidens tripartita	3106	NT
Blysmus compressus	434	LC
Bothriochloa ischaemum	161	LC
Brachypodium pinnatum	385	LC
Brachypodium sylvaticum	386	LC
Briza media	296	LC
Bromus benekenii	362	LC
Bromus commutatus	377	VU
Bromus erectus	363	NT
Bromus hordeaceus	370	LC
Bromus inermis	365	LC
Bryonia dioica	2884	LC
Buglossoides purpurocaerulea	2454	NT
Buphthalmum salicifolium	3119	LC
Bupleurum falcatum	2151	DD
Calamagrostis canescens	203	VU
Calamagrostis epigejos	201	LC
Calamagrostis varia	207	LC
Calamintha nepeta	2540	LC
Callitriche palustris	1970	LC
Calluna vulgaris	2271	LC
Caltha palustris	1125	LC

Calystegia sepium	2403	LC		Carex umbrosa	515	LC
Campanula glomerata	2919	VU		Carex vesicaria	524	NT
Campanula patula	2936	NT		Carex viridula	559	LC
Campanula persicifolia	2943	LC		Carex vulpinoidea	473	EN
Campanula rapunculoides	2922	LC		Carlina acaulis	2954	LC
Campanula rapunculus	2935	LC		Carlina simplex	2953	LC
Campanula rotundifolia	2931	LC		Carlina vulgaris	2951	LC
Campanula trachelium	2923	LC		Carpinus betulus	822	LC
Capsella bursa-pastoris	1276	LC		Carum carvi	2249	LC
Cardamine amara	1371	LC		Centaurea grinensis	3004	NT
Cardamine flexuosa	1364	LC		Centaurea jacea	3013	NT
Cardamine heptaphylla	1375	LC		Centaurea pannonica	3014	LC
Cardamine hirsuta	1363	LC		Centaurea scabiosa	3005	NT
Cardamine pentaphyllos	1377	LC		Centaurea stoebe	3017	EN
Cardamine pratensis	1369	LC		Centaurium erythraea	2361	LC
Carduus defloratus	2980	LC		Centaurium pulchellum	2362	VU
Carduus personata	2985	LC		Cephalanthera damasonium	730	LC
Carex acuta	507	LC		Cephalanthera longifolia	729	LC
Carex acutiformis	525	LC		Cephalanthera rubra	731	LC
Carex alba	528	LC		Cerastium arvense	1026	EN
Carex appropinquata	479	LC		Cerastium fontanum	1020	LC
Carex buxbaumii	500	EN		Cerastium glomeratum	1018	LC
Carex canescens	490	LC		Cerastium semidecandrum	1012	LC
Carex caryophyllea	520	LC		Cerastium vulgare	1019	LC
Carex davalliana	460	LC		Ceratophyllum demersum	1096	VU
Carex diandra	478	VU		Chaerophyllum hirsutum	2187	LC
Carex digitata	512	LC		Chenopodium bonus-henricus	913	LC
Carex dioica	461	NT		Chrysosplenium alternifolium	1526	LC
Carex distans	552	NT		Cichorium intybus	3232	LC
Carex disticha	492	NT		Cicuta virosa	2234	EN
Carex echinata	488	LC		Circaea lutetiana	2100	LC
Carex elata	503	LC		Cirsium acaule	2966	LC
Carex flacca	522	LC		Cirsium arvense	2961	LC
Carex flava	556	LC		Cirsium oleraceum	2971	LC
Carex hirta	508	LC		Cirsium palustre	2962	LC
Carex hostiana	553	LC		Cirsium spinosissimum	2972	LC
Carex humilis	511	LC		Cirsium tuberosum	2967	VU
Carex lasiocarpa	509	NT		Cirsium vulgare	2960	LC
Carex lepidocarpa	557	LC		Cladium mariscus	450	NT
Carex limosa	534	NT		Clematis vitalba	1153	LC
Carex montana	518	LC		Clinopodium vulgare	2535	LC
Carex muricata	475	LC		Colchicum autumnale	641	LC
Carex nigra	506	LC		Convallaria majalis	624	LC
Carex ornithopoda	513	LC		Convolvulus arvensis	2401	LC
Carex pallescens	555	LC		Cornus mas	2257	LC
Carex panicea	545	LC		Cornus sanguinea	2258	LC
Carex paniculata	480	LC		Coronilla coronata	1784	NT
Carex pendula	531	LC		Coronilla vaginalis	1782	LC
Carex pilulifera	516	LC		Corylus avellana	821	LC
Carex pseudocyperus	530	VU		Cotoneaster integerrimus	1653	LC
Carex pulicaris	458	NT		Cotoneaster tomentosus	1654	LC
Carex rostrata	523	LC		Crataegus laevigata	1656	LC
Carex spicata	474	LC		Crataegus monogyna	1657	LC
Carex sylvatica	537	LC		Crepis biennis	3318	LC
Carex tomentosa	519	LC		Crepis paludosa	3313	LC

Crepis praemorsa	3310	VU	Epipactis palustris	734	LC
Crepis pyrenaica	3309	LC	Epipactis purpurata	737	LC
Crepis taraxacifolia	3322	LC	Equisetum arvense	71	LC
Crepis vesicaria	3323	LC	Equisetum fluviatile	74	LC
Crocus albiflorus	701	LC	Equisetum hyemale	76	LC
Cucubalus baccifer	975	VU	Equisetum palustre	73	LC
Cuscuta epithymum	2406	LC	Equisetum sylvaticum	72	LC
Cynosurus cristatus	236	LC	Equisetum telmateia	69	LC
Cyperus flavescens	405	VU	Erigeron acer	3144	DD
Cyperus fuscus	406	VU	Erigeron annuus	3141	LC
Cypripedium calceolus	717	VU	Eriophorum angustifolium	432	LC
Cystopteris fragilis	35	LC	Eriophorum gracile	433	EN
Dactylis glomerata	294	LC	Eriophorum latifolium	431	LC
Dactylorhiza fuchsii	778	LC	Erophila verna	1309	LC
Dactylorhiza incarnata	775	NT	Erucastrum gallicum	1350	NT
Dactylorhiza maculata	777	CR	Euonymus europaeus	1981	LC
Dactylorhiza majalis	781	LC	Eupatorium cannabinum	3034	LC
Dactylorhiza traunsteineri	779	NT	Euphorbia amygdaloides	1959	LC
Danthonia decumbens	246	LC	Euphorbia cyparissias	1960	LC
Daphne laureola	2084	LC	Euphorbia palustris	1952	VU
Daphne mezereum	2085	LC	Euphorbia platyphyllos	1953	LC
Daucus carota	2139	LC	Euphorbia verrucosa	1957	LC
Deschampsia cespitosa	272	LC	Euphrasia rostkoviana	2717	LC
Dianthus armeria	990	NT	Fagus sylvatica	832	LC
Dianthus carthusianorum	989	LC	Festuca arundinacea	339	DD
Dianthus superbus	985	LC	Festuca laevigata	354	LC
Digitalis grandiflora	2671	LC	Festuca ovina	355	LC
Digitalis lutea	2670	LC	Festuca pratensis	338	LC
Dipsacus pilosus	2862	VU	Festuca rubra	341	LC
Draba aizoides	1308	LC	Filipendula ulmaria	1563	LC
Draba muralis	1298	VU	Filipendula vulgaris	1564	VU
Drosera anglica	1444	VU	Fragaria vesca	1568	LC
Drosera rotundifolia	1443	NT	Frangula alnus	2001	LC
Dryopteris carthusiana	27	LC	Fraxinus excelsior	2354	LC
Dryopteris filix-mas	22	LC	Galanthus nivalis	696	NT
Echium vulgare	2419	LC	Galeopsis angustifolia	2496	NT
Eleocharis acicularis	422	VU	Galeopsis bifida	2502	VU
Eleocharis austriaca	417	NT	Galeopsis tetrahit	2501	LC
Eleocharis palustris	416	LC	Galium album	2806	LC
Eleocharis quinqueflora	420	LC	Galium anisophyllon	2797	LC
Eleocharis uniglumis	419	NT	Galium aparine	2814	LC
Elodea canadensis	154	LC	Galium boreale	2783	LC
Elymus repens	390	LC	Galium elongatum	2788	NT
Epilobium angustifolium	2105	LC	Galium mollugo	2805	LC
Epilobium collinum	2110	LC	Galium odoratum	2785	LC
Epilobium dodonaei	2106	LC	Galium palustre	2787	LC
Epilobium hirsutum	2109	LC	Galium pumilum	2795	LC
Epilobium montanum	2111	LC	Galium sylvaticum	2812	LC
Epilobium palustre	2114	LC	Galium uliginosum	2790	LC
Epilobium parviflorum	2108	LC	Galium verum	2803	LC
Epilobium tetragonum	2122	NT	Genista germanica	1682	LC
Epipactis atrorubens	735	LC	Genista tinctoria	1681	LC
Epipactis helleborine	738	LC	Gentiana asclepiadea	2368	LC
Epipactis microphylla	736	NT	Gentiana campestris	2386	LC
Epipactis muelleri	739	LC	Gentiana ciliata	2383	LC

Gentiana germanica	2388	NT		Hypochaeris radicata	3256	LC
Gentiana pneumonanthe	2367	VU		Ilex aquifolium	1979	LC
Geranium dissectum	1887	LC		Impatiens glandulifera	1993	LC
Geranium molle	1890	LC		Impatiens noli-tangere	1991	LC
Geranium palustre	1900	NT		Impatiens parviflora	1992	LC
Geranium pyrenaicum	1891	LC		Inula conyzae	3124	LC
Geranium robertianum	1883	LC		Inula graveolens	3122	DD
Geranium sanguineum	1895	LC		Inula salicina	3131	NT
Geum rivale	1556	LC		Iris pseudacorus	706	LC
Geum urbanum	1557	LC		Iris sibirica	707	VU
Glechoma hederacea	2544	LC		Isolepis setacea	443	VU
Globularia bisnagarica	2755	LC		Juglans regia	820	LC
Globularia cordifolia	2756	LC		Juncus acutiflorus	600	NT
Glyceria fluitans	303	LC		Juncus articulatus	599	LC
Glyceria maxima	301	VU		Juncus bufonius	581	LC
Glyceria notata	304	LC		Juncus conglomeratus	578	LC
Goodyera repens	741	LC		Juncus effusus	579	LC
Groenlandia densa	124	NT		Juncus fuscoater	598	LC
Gymnadenia conopsea	756	LC		Juncus inflexus	577	LC
Gymnadenia odoratissima	757	LC		Juncus subnodulosus	597	LC
Gymnocarpium dryopteris	18	LC		Juniperus communis	101	LC
Hedera helix	2130	LC		Kernera saxatilis	1297	LC
Helianthemum grandiflorum	2046	LC		Knautia arvensis	2868	LC
Helianthemum nummularium	2044	LC		Knautia dipsacifolia	2867	LC
Helianthemum ovatum	2045	LC		Koeleria macrantha	288	LC
Helictotrichon pubescens	260	LC		Koeleria pyramidata	286	LC
Helleborus foetidus	1123	LC		Lactuca perennis	3283	LC
Hemerocallis fulva	634	LC		Lactuca serriola	3288	LC
Hepatica nobilis	1149	LC		Lamium galeobdolon	2508	LC
Heracleum mantegazzianum	2211	LC		Lamium maculatum	2512	LC
Heracleum sphondylium	2207	NT		Lamium purpureum	2513	LC
Hieracium amplexicaule	3347	LC		Larix decidua	100	LC
Hieracium cymosum	3328	NT		Laserpitium latifolium	2162	LC
Hieracium humile	3348	LC		Laserpitium siler	2160	LC
Hieracium lactucella	3333	LC		Lathraea squamaria	2725	LC
Hieracium murorum	3350	LC		Lathyrus linifolius	1868	LC
Hieracium pilosella	3335	LC		Lathyrus niger	1874	LC
Hieracium piloselloides	3329	LC		Lathyrus pratensis	1861	LC
Hieracium umbellatum	3361	LC		Lathyrus sylvestris	1865	LC
Himantoglossum hircinum	750	VU		Lathyrus vernus	1872	NT
Hippocrepis comosa	1786	LC		Leersia oryzoides	230	EN
Hippocrepis emerus	1780	LC		Legousia speculum-veneris	2909	VU
Hippuris vulgaris	2129	NT		Lemna minor	569	LC
Holcus lanatus	238	LC		Lemna trisulca	567	NT
Holcus mollis	239	LC		Leontodon autumnalis	3265	LC
Hordelymus europaeus	402	LC		Leontodon hispidus	3263	LC
Hottonia palustris	2290	EN		Leucanthemum adustum	3203	LC
Humulus lupulus	849	LC		Leucanthemum ircutianum	3199	LC
Hydrocharis morsus-ranae	153	EN		Leucanthemum ircutianum	3197	LC
Hydrocotyle vulgaris	2131	VU		Leucojum vernum	697	LC
Hypericum desetangsii	2027	LC		Ligustrum vulgare	2351	LC
Hypericum dubium	2026	LC		Lilium bulbiferum	652	NT
Hypericum maculatum	2025	LC		Lilium martagon	650	LC
Hypericum montanum	2031	LC		Limodorum abortivum	719	NT
Hypericum perforatum	2023	LC		Linaria vulgaris	2650	LC

Linum catharticum	1914	LC		Myosotis arvensis	2451	LC
Linum tenuifolium	1916	LC		Myosotis scorpioides	2441	LC
Liparis loeselii	744	VU		Myosotis sylvatica	2445	LC
Listera ovata	747	LC		Myosoton aquaticum	999	LC
Lithospermum officinale	2455	NT		Myriophyllum spicatum	2126	NT
Lolium multiflorum	381	LC		Nasturtium officinale	1378	LC
Lolium perenne	380	LC		Neottia nidus-avis	721	LC
Lonicera alpigena	2827	LC		Nuphar lutea	1092	LC
Lonicera nigra	2830	LC		Nymphaea alba	1090	NT
Lonicera periclymenum	2832	NT		Nymphoides peltata	2358	VU
Lonicera xylosteum	2829	LC		Onobrychis viciifolia	1817	LC
Lotus corniculatus	1763	LC		Ononis repens	1704	LC
Lotus maritimus	1767	LC		Ononis rotundifolia	1700	LC
Lotus pedunculatus	1762	LC		Ononis spinosa	1703	NT
Lunaria rediviva	1290	LC		Onopordum acanthium	2987	VU
Luzula campestris	613	LC		Ophioglossum vulgatum	59	VU
Luzula multiflora	614	LC		Ophrys apifera	726	EN
Luzula pilosa	603	LC		Ophrys holosericea	724	EN
Luzula sylvatica	609	LC		Ophrys insectifera	727	NT
Lycopodium annotinum	80	LC		Ophrys sphegodes	728	EN
Lycopus europaeus	2558	EN		Orchis mascula	773	LC
Lysimachia nemorum	2331	LC		Orchis militaris	767	NT
Lysimachia nummularia	2332	LC		Orchis morio	762	NT
Lysimachia thyrsiflora	2335	VU		Orchis pallens	769	NT
Lysimachia vulgaris	2334	LC		Orchis purpurea	766	VU
Lythrum salicaria	2091	LC		Orchis ustulata	764	NT
Malus sylvestris	1666	NT		Origanum majorana	2550	DD
Malva moschata	2008	LC		Origanum vulgare	2549	LC
Medicago falcata	1757	LC		Ornithogalum umbellatum	656	LC
Medicago lupulina	1749	LC		Orobanche caryophyllacea	2733	LC
Medicago minima	1752	LC		Orobanche elatior	2738	EN
Medicago sativa	1756	LC		Orobanche minor	2737	LC
Melampyrum arvense	2701	VU		Orobanche reticulata	2732	LC
Melampyrum pratense	2704	LC		Orobanche teucrii	2743	LC
Melampyrum sylvaticum	2705	LC		Orthilia secunda	2261	LC
Melica ciliata	232	LC		Oxalis acetosella	1909	LC
Melica nutans	234	LC		Oxalis stricta	1912	NA
Melica uniflora	235	LC		Papaver rhoeas	1223	LC
Melilotus albus	1740	LC		Paris quadrifolia	622	LC
Melilotus officinalis	1741	LC		Parnassia palustris	1525	LC
Melittis melissophyllum	2547	LC		Pastinaca sativa	2212	LC
Mentha aquatica	2563	LC		Pedicularis palustris	2676	LC
Mentha arvensis	2562	LC		Pedicularis sylvatica	2677	VU
Mentha longifolia	2565	LC		Petasites albus	3041	LC
Menyanthes trifoliata	2357	LC		Petrorhagia prolifera	984	LC
Mercurialis perennis	1940	LC		Peucedanum cervaria	2218	LC
Milium effusum	209	LC		Peucedanum oreoselinum	2217	LC
Moehringia muscosa	1045	LC		Peucedanum palustre	2225	NT
Moehringia trinervia	1041	LC		Phalaris arundinacea	178	LC
Molinia arundinacea	282	LC		Phleum bertolonii	223	LC
Molinia caerulea	281	LC		Phleum pratense	222	LC
Monotropa hypopitys	2267	LC		Phragmites australis	242	LC
Muscari comosum	689	LC		Phyllitis scolopendrium	56	LC
Muscari racemosum	687	NT		Phyteuma orbiculare	2899	LC
Mycelis muralis	3290	LC		Phyteuma spicatum	2906	LC

Picea abies	92	LC
Picris hieracioides	3269	LC
Picris hieracioides	2369	LC
Pimpinella major	2245	LC
Pimpinella saxifraga	2246	LC
Pinguicula vulgaris	2747	LC
Pinus sylvestris	96	LC
Plantago atrata	2764	LC
Plantago lanceolata	2761	LC
Plantago major	2758	LC
Plantago media	2760	LC
Platanthera bifolia	753	LC
Platanthera chlorantha	754	LC
Poa angustifolia	325	LC
Poa annua	307	LC
Poa bulbosa	311	LC
Poa palustris	320	LC
Poa pratensis	324	LC
Poa trivialis	309	VU
Polygala amara	1351	DD
Polygala amarella	1932	LC
Polygala comosa	1936	LC
Polygala vulgaris	1934	NT
Polygonatum multiflorum	626	LC
Polygonatum odoratum	627	LC
Polygonatum verticillatum	625	LC
Polygonum amphibium	906	NT
Polygonum aviculare	899	LC
Polygonum bistorta	897	LC
Polystichum aculeatum	14	LC
Populus alba	784	LC
Populus nigra	782	LC
Populus tremula	783	LC
Potamogeton crispus	123	LC
Potamogeton gramineus	126	EN
Potamogeton natans	116	LC
Potamogeton perfoliatus	121	LC
Potentilla anserina	1580	LC
Potentilla erecta	1606	LC
Potentilla neumanniana	1601	LC
Potentilla palustris	1566	LC
Potentilla recta	1587	LC
Potentilla reptans	1605	LC
Potentilla sterilis	1577	LC
Prenanthes purpurea	3282	LC
Primula auricula	2304	LC
Primula elatior	2298	LC
Primula farinosa	2302	LC
Primula veris	2300	LC
Primula veris ssp columnae	2301	LC
Prunella grandiflora	2495	LC
Prunella laciniata	2493	EN
Prunella vulgaris	2494	LC
Prunus avium	1648	LC
Prunus domestica	1647	LC

Prunus padus	1637	NT
Prunus spinosa	1644	LC
Pteridium aquilinum	8	LC
Pulicaria dysenterica	3134	LC
Pulmonaria officinalis	2434	LC
Pulsatilla vulgaris	1146	EN
Pyrola rotundifolia	2264	LC
Pyrus pyraster	1665	LC
Quercus petraea	835	LC
Quercus pubescens	836	LC
Quercus robur	834	LC
Ranunculus aconitifolius	1171	LC
Ranunculus acris	1207	LC
Ranunculus acris ssp friesianus	1206	LC
Ranunculus auricomus	1185	LC
Ranunculus bulbosus	1192	LC
Ranunculus flammula	1178	NT
Ranunculus fluitans	1161	NT
Ranunculus lingua	1177	VU
Ranunculus repens	1193	LC
Ranunculus trichophyllus	1162	NT
Ranunculus tuberosus	1195	LC
Raphanus raphanistrum	1335	LC
Reseda lutea	1439	LC
Reseda luteola	1440	VU
Reynoutria japonica	895	LC
Rhamnus alpina	1998	LC
Rhamnus cathartica	1996	LC
Rhinanthus alectorolophus	2694	LC
Rhinanthus angustifolius	2695	VU
Rhinanthus glacialis	2697	LC
Rhinanthus minor	2699	LC
Rhynchospora alba	451	NT
Ribes rubrum	1531	LC
Ribes uva-crispa	1528	LC
Robinia pseudoacacia	1769	LC
Rorippa amphibia	1385	VU
Rorippa islandica	1381	NT
Rosa arvensis	1614	LC
Rosa canina	1630	LC
Rosa glauca	1618	LC
Rosa jundzillii	1622	NT
Rosa pendulina	1616	LC
Rosa spinosissima	1615	LC
Rosa tomentella	1632	VU
Rosa tomentosa	1624	LC
Rubus caesius	1610	LC
Rubus fruticosus	1611	LC
Rubus idaeus	1609	LC
Rubus saxatilis	1608	LC
Rumex acetosa	876	LC
Rumex acetosella	871	LC
Rumex hydrolapathum	883	EN
Rumex obtusifolius	888	LC
Sagittaria latifolia	143	VU

Sagittaria sagittifolia	142	EN
Salix alba	793	LC
Salix appendiculata	814	LC
Salix aurita	817	LC
Salix caprea	815	LC
Salix cinerea	818	LC
Salix elaeagnos	791	LC
Salix myrsinifolia	800	LC
Salix purpurea	789	LC
Salix repens	796	NT
Salix triandra	805	LC
Salix viminalis	792	LC
Salvia glutinosa	2473	LC
Salvia pratensis	2477	LC
Sambucus nigra	2821	LC
Sambucus racemosa	2822	LC
Sanguisorba minor	1538	NT
Sanguisorba officinalis	1536	LC
Sanicula europaea	2138	LC
Saponaria officinalis	979	LC
Saxifraga paniculata	1487	LC
Saxifraga tridactylites	1522	LC
Scabiosa columbaria	2879	LC
Schoenoplectus lacustris	448	LC
Schoenoplectus tabernaemontani	449	VU
Schoenus ferrugineus	414	NT
Schoenus nigricans	413	NT
Scilla bifolia	653	LC
Scirpus sylvaticus	436	LC
Scrophularia canina	2666	LC
Scrophularia nodosa	2663	LC
Scutellaria galericulata	2482	LC
Securigera varia	1785	LC
Sedum acre	1476	LC
Sedum album	1474	LC
Sedum rubens	1478	VU
Sedum sexangulare	1477	LC
Sedum telephium	1460	LC
Selinum carvifolia	2202	VU
Senecio aquaticus	3097	NT
Senecio erucifolius	3094	LC
Senecio jacobaea	3095	LC
Senecio paludosus	3081	NT
Serratula tinctoria	2994	NT
Seseli libanotis	2175	LC
Sesleria caerulea	251	LC
Silaum silaus	2184	NT
Silene dioica	954	LC
Silene flos-cuculi	953	LC
Silene nutans	959	NT
Silene pratensis	955	LC
Silene vulgaris	965	NT
Sinapis arvensis	1336	LC
Sium latifolium	2244	CR

Solanum dulcamara	2569	LC
Solidago canadensis	3136	LC
Solidago virgaurea	3138	LC
Sonchus asper	3296	LC
Sonchus oleraceus	3295	LC
Sorbus aria	1663	LC
Sorbus aucuparia	1659	LC
Sorbus mougeotii	1664	LC
Sorbus torminalis	1661	LC
Sparganium emersum	113	VU
Sparganium erectum	110	EN
Spiranthes aestivalis	733	VU
Spiranthes spiralis	732	NT
Stachys annua	2517	VU
Stachys officinalis	2527	VU
Stachys palustris	2520	NT
Stachys recta	2518	VU
Stachys sylvatica	2521	LC
Stellaria graminea	1007	LC
Stellaria media	1002	LC
Stratiotes aloides	159	VU
Succisa pratensis	2864	LC
Symphytum officinale	2432	LC
Tamus communis	690	LC
Tanacetum corymbosum	3191	NT
Tanacetum vulgare	3193	LC
Taraxacum officinale	3272	LC
Taraxacum palustre	3271	LC
Taxus baccata	90	LC
Teucrium chamaedrys	2468	LC
Teucrium montanum	2466	LC
Teucrium scordium	2469	EN
Teucrium scorodonia	2471	LC
Thalictrum aquilegiifolium	1126	LC
Thalictrum flavum	1133	VU
Thalictrum minus	1128	NT
Thelypteris palustris	21	VU
Thesium alpinum	860	LC
Thesium bavarum	858	NT
Thesium linophyllon	856	VU
Thesium pyrenaicum	861	LC
Thlaspi montanum	1269	LC
Thlaspi perfoliatum	1268	LC
Thymus oenipontanus	2555	LC
Thymus praecox ssp. polytrichus	2554	LC
Thymus pulegioides	2556	LC
Thymus pulegioides ssp. carniolicus	2557	LC
Tilia cordata	2005	LC
Tilia platyphyllos	2006	LC
Tofieldia calyculata	618	LC
Torilis japonica	2145	LC
Tragopogon pratensis	3240	LC
Tragopogon pratensis ssp.	3239	LC

orientalis		
Trichophorum cespitosum	426	LC
Trifolium aureum	1708	NT
Trifolium campestre	1710	LC
Trifolium dubium	1711	LC
Trifolium fragiferum	1714	VU
Trifolium medium	1735	LC
Trifolium montanum	1718	LC
Trifolium pratense	1737	LC
Trifolium repens	1723	VU
Trifolium rubens	1730	NT
Trifolium thalii	1724	LC
Triglochin palustris	141	LC
Trisetum flavescens	277	LC
Trollius europaeus	1120	LC
Tussilago farfara	3043	LC
Typha angustifolia	108	NT
Typha latifolia	106	LC
Typha minima	109	EN
Typha shuttleworthii	107	VU
Ulmus glabra	841	LC
Ulmus minor	840	NT
Urtica dioica	851	LC
Utricularia minor	2751	VU
Vaccinium myrtillus	2281	LC
Vaccinium oxycoccos	2277	NT
Vaccinium uliginosum	2283	LC
Vaccinium vitis-idaea	2280	LC
Valeriana dioica	2846	LC
Valeriana montana	2844	LC
Valeriana officinalis	2848	LC
Valeriana repens	2852	LC
Valeriana wallrothii	2850	VU
Valerianella locusta	2855	LC
Verbascum lychnitis	2595	LC
Verbascum nigrum	2597	LC
Verbascum thapsus	2589	LC
Verbena officinalis	2461	LC
Veronica anagallis-aquatica	2603	LC
Veronica arvensis	2624	LC
Veronica beccabunga	2602	LC
Veronica chamaedrys	2614	LC
Veronica filiformis	2634	LC
Veronica persica	2633	LC
Veronica polita	2630	LC
Veronica scutellata	2611	VU
Veronica serpyllifolia	2621	LC
Veronica spicata	2638	LC
Veronica teucrium	2606	LC
Viburnum lantana	2824	LC
Viburnum opulus	2825	LC
Vicia cracca	1825	NT
Vicia cracca ssp. tenuifolia	1827	NT
Vicia dumetorum	1832	NT
Vicia hirsuta	1821	LC

Vicia sativa	1840	LC
Vicia sativa ssp. nigra	1844	LC
Vicia sepium	1835	LC
Vicia sylvatica	1830	LC
Vicia tetrasperma	1822	NT
Vicia villosa	1829	VU
Vinca minor	2397	LC
Vincetoxicum hirundinaria	2399	LC
Viola alba	2069	VU
Viola canina	2076	NT
Viola hirta	2065	LC
Viola palustris	2063	LC
Viola reichenbachiana	2074	LC
Viola tricolor	2053	LC
Zannichellia palustris	135	VU

Animal Species	ID	Red List Status
Accipiter nisus	96716	LC
Acrocephalus arundinaceus	97319	NT
Acrocephalus palustris	97317	LC
Acrocephalus schoenobaenus	97314	NE
Acrocephalus scirpaceus	97318	LC
Acroloxus lacustris	421525	LC
Acronicta aceris	449422	NA
Acronicta psi	449420	NA
Actitis hypoleucos	96856	EN
Aculepeira ceropegia	348377	NA
Adscita statices	440602	NA
Aegithalos caudatus	97278	LC
Aeshna affinis	214299	NE
Aeshna cyanea	214303	LC
Aeshna isosceles	214305	LC
Aeshna mixta	214308	LC
Agapanthia violacea	114168	NA
Aglais io	441678	NA
Aglais urticae	441679	NA
Aglia tau	443627	NA
Agrius convolvuli	443845	NA
Agrotis ipsilon	448502	NA
Alauda arvensis	97404	NT
Alcedo atthis	97082	VU
Alsophila aceraria	446361	NA
Alytes obstetricans	177873	EN
Ampedus sinuatus	235549	NA
Anas crecca	96500	VU
Anas platyrhynchos	96502	LC
Anas strepera	96498	EN
Anax imperator	214314	LC
Anax parthenope	214316	LC
Anguis fragilis	214587	LC
Anisus leucostoma	430524	LC
Anthocharis cardamines	440867	NA
Anthus trivialis	97417	LC
Apatura iris	441662	VU
Aplocera plagiata	444530	NA
Aporia crataegi	440897	VU
Araschnia levana	441672	NA
Ardea cinerea	96658	LC
Argiope bruennichi	348443	NE
Argynnis paphia	441744	NA
Aricia agestis	441006	VU
Asio otus	97051	NT
Auchmis detersa	447366	NA
Autographa gamma	449605	NA
Aythya ferina	96513	EN
Aythya fuligula	96519	VU
Boloria dia	441716	EN
Boloria euphrosyne	441700	NA
Bombina variegata	177877	EN

Animal Species	ID	Red List Status
Bombus sylvarum	231919	NA
Brachytron pratense	214318	LC
Brenthis ino	441735	VU
Bufo bufo	177878	VU
Bufo calamita	177879	EN
Buteo buteo	96722	LC
Calliptamus italicus	402751	VU
Calliteara pudibunda	447103	NA
Callophrys rubi	441177	VU
Calopteryx splendens	214240	LC
Carduelis cannabina	97472	NT
Carduelis chloris	97465	LC
Carduelis spinus	97471	LC
Carterocephalus palaemon	440735	NA
Castor fiber	305558	CR
Catocala fulminea	446652	NA
Catocala promissa	446678	NA
Cerambyx scopolii	114746	NA
Cercion lindenii	214258	NT
Certhia brachydactyla	97261	LC
Chalicodoma parietina	231979	VU
Charadrius dubius	96870	EN
Chorthippus albomarginatus	402549	LC
Chorthippus mollis	402587	NT
Chrysochraon dispar	402547	NT
Cicadetta montana	239538	NA
Cicindela germanica	387305	EN
Ciconia ciconia	96674	VU
Cinclus cinclus	97156	LC
Coccothraustes coccothraustes	97498	LC
Coenagrion puella	214272	LC
Coenonympha arcania	441264	NT
Coenonympha pamphilus	441285	NA
Coenonympha tullia	441253	EN
Colias croceus	440816	NA
Colias hyale	440826	NA
Columba oenas	97003	LC
Columba palumbus	97005	LC
Conocephalus fuscus	403179	VU
Cordulegaster bidentata	214348	NT
Cordulegaster boltonii	214349	LC
Cordulia aenea	214357	LC
Coronella austriaca	214732	VU
Corvus corax	97146	LC
Corvus corone	97143	LC
Crocothemis erythraea	214375	LC
Cucullia umbratica	449318	NA
Cuculus canorus	97028	NT
Cupido argiades	441120	CR
Cupido minimus	441111	VU
Cyaniris semiargus	440997	NA
Decticus verrucivorus	403450	NT
Deilephila porcellus	443741	NA
Dendrocopos major	97099	LC
Dendrocopos medius	97101	NT

68

Dendrocopos minor	97103	LC
Diachrysia chrysitis	449550	NA
Dolomedes fimbriatus	352481	NA
Dryocopus martius	97095	LC
Ectropis crepuscularia	445706	NA
Emberiza cirlus	97524	NT
Emberiza citrinella	97523	LC
Emberiza schoeniclus	97536	VU
Erebia aethiops	441347	VU
Erebia ligea	441321	NA
Erinaceus europaeus	305459	LC
Eriogaster lanestris	443530	NA
Erithacus rubecula	97170	LC
Erynnis tages	440801	NA
Erythromma najas	214278	LC
Erythromma viridulum	214279	LC
Euclidia glyphica	446607	NA
Euconulus alderi	425745	NT
Euthystira brachyptera	402521	LC
Euxoa nigrofusca	448566	NA
Falco subbuteo	96742	NT
Falco tinnunculus	96738	NT
Favonius quercus	441154	NA
Ficedula hypoleuca	97165	LC
Formica polyctena	81321	NT
Formica rufa	81325	NT
Fringilla coelebs	97456	LC
Fringilla montifringilla	97459	NA
Fulica atra	96786	LC
Gallinula chloropus	96779	LC
Garrulus glandarius	97127	LC
Gerris lacustris	450573	NA
Glis glis	305715	LC
Gomphus vulgatissimus	214334	NT
Gonepteryx rhamni	440831	NA
Gryllus campestris	402997	LC
Gyraulus laevis	430518	VU
Hadena compta	448102	NA
Haitia acuta	430438	NE
Hamearis lucina	440909	VU
Hemaris fuciformis	443647	NA
Hemaris tityus	443640	NA
Hemianax ephippiger	214326	NE
Hesperia comma	440721	NA
Hippeutis complanatus	430484	LC
Hippolais icterina	97334	VU
Hyla arborea	177887	EN
Hyles euphorbiae	443766	NA
Hyles gallii	443793	NA
Iphiclides podalirius	440667	EN
Ischnura elegans	214281	LC
Ischnura pumilio	214286	LC
Issoria lathonia	441740	NA
Ixobrychus minutus	96634	EN
Jordanita globulariae	440643	NA

Jordanita notata	440630	NA
Jynx torquilla	97090	NT
Lacerta agilis	214645	VU
Lanius collurio	97116	LC
Lanius senator	97121	CR
Larus ridibundus	96929	EN
Lasiocampa trifolii	443542	NA
Lasiommata megera	441226	NA
Leptidea sinapis	440900	NA
Leptophyes punctatissima	403278	LC
Lepus europaeus	305525	VU
Lestes viridis	214252	LC
Leucorrhinia caudalis	214380	CR
Leucorrhinia pectoralis	214382	CR
Libelloides coccajus	85237	VU
Libellula depressa	214384	LC
Libellula fulva	214385	LC
Libellula quadrimaculata	214386	LC
Limenitis reducta	441651	EN
Locustella luscinioides	97309	NT
Locustella naevia	97307	NT
Lopinga achine	441223	EN
Lucanus cervus	123296	NA
Luscinia megarhynchos	97174	NT
Lycaena dispar	441190	NA
Lycaena phlaeas	441185	NA
Lycaena tityrus	441198	NA
Lycaena virgaureae	441193	VU
Macroglossum stellatarum	443702	NA
Maniola jurtina	441303	NA
Mecostethus parapleurus	402378	LC
Melanargia galathea	441445	NA
Melanchra persicariae	448252	NA
Melitaea athalia	441611	VU
Melitaea diamina	441597	VU
Melitaea didyma	441594	VU
Melitaea parthenoides	441604	EN
Melitta nigricans	232909	EN
Mergus merganser	96551	VU
Metrioptera bicolor	403499	VU
Metrioptera brachyptera	403500	NT
Microlinyphia impigra	350719	NA
Miliaria calandra	97541	VU
Milvus migrans	96688	LC
Milvus milvus	96689	LC
Minucia lunaris	446631	NA
Motacilla alba	97427	LC
Motacilla cinerea	97426	LC
Muscardinus avellanarius	305717	VU
Muscicapa striata	97159	LC
Musculium lacustre	337909	LC
Mustela erminea	305315	LC
Mustela nivalis	305324	VU
Myotis bechsteinii	305408	NT
Myotis daubentonii	305417	VU

Myotis myotis	305419	EN		Polygonia c-album	441674	NA
Myotis nattereri	305421	NT		Polyommatus bellargus	440948	NA
Natrix natrix	214751	EN		Polyommatus coridon	440950	VU
Noctua fimbriata	448845	NA		Polyommatus icarus	440934	NA
Noctua pronuba	448843	NA		Polyommatus thersites	440933	VU
Notonecta glauca	450468	NA		Pontia daplidice	440877	EN
Nyctalus noctula	305429	VU		Prunella modularis	97429	LC
Nymphalis antiopa	441666	VU		Pseudochorthippus montanus	402589	VU
Nymphalis polychloros	441667	VU		Psophus stridulus	402353	VU
Odezia atrata	444521	NA		Pteronemobius heydenii	402927	VU
Oecanthus pellucens	402913	LC		Pupilla muscorum	430797	NA
Oedipoda caerulescens	402365	NT		Pyrgus malvae	440751	VU
Omocestus rufipes	402488	NT		Pyrrhosoma nymphula	214292	LC
Omocestus viridulus	402492	LC		Pyrrhula pyrrhula	97495	LC
Onychogomphus forcipatus	214340	EN		Rallus aquaticus	96767	LC
Opisthograptis luteolata	446005	NA		Rana esculenta	177908	NT
Orgyia antiqua	447075	NA		Rana lessonae	177916	NT
Oriolus oriolus	97124	LC		Rana ridibunda	177919	NE
Orthetrum albistylum	214388	NA		Rana temporaria	177921	LC
Orthetrum brunneum	214389	LC		Regulus ignicapillus	97297	LC
Orthetrum cancellatum	214390	LC		Regulus regulus	97294	LC
Orthetrum coerulescens	214392	NT		Rhizotrogus aestivus	257568	NA
Papilio machaon	440671	NA		Rhodostrophia vibicaria	444265	NA
Pararge aegeria	441236	NA		Riparia riparia	97282	VU
Parnassius apollo	440702	VU		Ruspolia nitidula	403186	NT
Parus ater	97272	LC		Salamandra salamandra	177846	VU
Parus caeruleus	97274	LC		Satyrium pruni	441169	CR
Parus major	97276	LC		Saxicola rubetra	97189	VU
Parus palustris	97267	LC		Saxicola rubicola	97192	NT
Passer domesticus	97437	LC		Scolopax rusticola	96827	VU
Passer montanus	97441	LC		Segmentina nitida	430462	VU
Phalera bucephala	446378	NA		Serinus serinus	97462	LC
Phaneroptera falcata	403293	VU		Siona lineata	446008	NA
Phengaris arion	441074	VU		Sitta europaea	97254	LC
Phoenicurus ochruros	97184	LC		Somatochlora flavomaculata	214366	LC
Phoenicurus phoenicurus	97185	NT		Somatochlora metallica	214368	LC
Phragmatobia fuliginosa	447012	NA		Sphingonotus caerulans	402321	VU
Phragmatobia luctifera	447014	NA		Sphinx ligustri	443872	NA
Phylloscopus bonelli	97367	LC		Spialia sertorius	440779	NA
Phylloscopus collybita	97373	LC		Spilosoma lubricipeda	446997	NA
Phylloscopus trochilus	97375	VU		Spilosoma lutea	446996	NA
Pica pica	97133	LC		Stenobothrus lineatus	402447	LC
Picus canus	97092	VU		Sterna hirundo	96959	NT
Picus viridis	97093	LC		Stethophyma grossum	402319	VU
Pieris brassicae	440880	NA		Streptopelia turtur	97013	NT
Pieris napi	440887	NA		Strix aluco	97047	LC
Pieris rapae	440885	NA		Sturnus vulgaris	97244	LC
Plagodis dolabraria	446227	NA		Sylvia atricapilla	97352	LC
Planorbis carinatus	430469	LC		Sylvia borin	97351	NT
Platycleis albopunctata	403578	NT		Sylvia communis	97350	NT
Plebejus argus	441032	VU		Sympecma fusca	214254	LC
Plebejus idas	441036	VU		Sympetrum depressiusculum	214400	VU
Plecotus auritus	305438	VU		Sympetrum flaveolum	214401	EN
Podarcis muralis	214670	LC		Sympetrum fonscolombii	214402	NE
Podiceps cristatus	96561	LC		Sympetrum pedemontanum	214406	CR

70

Sympetrum sanguineum	214407	LC		Turdus philomelos	97234	LC
Sympetrum striolatum	214409	LC		Turdus pilaris	97228	VU
Sympetrum vulgatum	214410	LC		Tyto alba	97035	NT
Tachybaptus ruficollis	96559	VU		Vallonia excentrica	431110	NA
Tetragnatha extensa	353100	NA		Vanellus vanellus	96893	CR
Tetralonia malvae	233474	NA		Vanessa atalanta	441684	NA
Tetrix bipunctata	402277	NT		Vanessa cardui	441686	NA
Thecla betulae	441157	NA		Vertigo moulinsiana	431209	EN
Thera juniperata	444646	NA		Vipera aspis	214763	CR
Thymelicus sylvestris	440726	NA		Vipera berus	214764	EN
Trachys troglodytiformis	102594	NA		Vulpes vulpes	305294	LC
Trichodes alvearius	187146	NA		Xerolenta obvia	428799	NT
Tringa ochropus	96850	LC		Xestia c-nigrum	448764	NA
Triphosa dubitata	445159	NA		Xestia triangulum	448766	NA
Triturus alpestris	177854	LC		Xylena exsoleta	447671	NA
Triturus cristatus	177858	EN		Zebrina detrita	425370	VU
Triturus helveticus	177861	VU		Zootoca vivipara	214687	LC
Triturus vulgaris	177866	EN		Zygaena carniolica	440326	NA
Troglodytes troglodytes	97263	LC		Zygaena filipendulae	440536	NA
Truncatellina cylindrica	431271	VU		Zygaena loti	440360	NA
Turdus merula	97226	LC		Zygaena purpuralis	440279	NA

71

 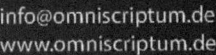